W9-BCY-753

GIROLAMO SACCHERI'S

EUCLIDES VINDICATUS

EDITED AND TRANSLATED BY

GEORGE BRUCE HALSTED

A.M., PRINCETON, PH.D., JOHNS HOPKINS

CHELSEA PUBLISHING COMPANY

NEW YORK, N. Y.

SECOND EDITION

Original Latin edition published at Milan, 1733

First English Edition (with the original Latin on facing pages) was published at Chicago in 1920

The present Second English Edition (with the original Latin on facing pages) and with added notes by Paul Stäckel and Friedrich Engel, translated from the German by F. Steinhardt, is published at New York, N.Y. in 1986. It is printed on 'long-life' acid-free paper

Library of Congress Catalog Card Number 79-16890
International Standard Book Number 0-8284-0289-2

QA 685.S313 516'.9

Printed in the United States of America

TABLE OF CONTENTS.

PART II.

PREFACE TO THE SECOND EDITION

In this book Girolamo Saccheri set forth in 1733, for the first time ever, what amounts to the axiom systems of non-Euclidean geometry. It is in this book that, for the first time in history, theorem after theorem of (hyperbolic) non-Euclidean geometry is stated and proved. The system of axioms of non-Euclidean geometry, the important theorems of what was to be a new discipline, and the proofs of those theorems—all the purposeful, deliberate creation of an acute mind and a driving will.

Why—one cannot help asking—did Saccheri not evaluate correctly what he had achieved, why did he not claim credit for the discovery of non-Euclidean geometry?

An almost unavoidable question, but one too narrowly put.

For while it is indeed true, as is well known, that Saccheri's book is a lost masterpiece, that it was forgotten for a century and a half, until it was dramatically rediscovered in 1889, this statement of truth slurs over several facts. First, the book was not ignored *initially;* indeed, on publication it created something of a sensation. It was read and examined by some leading mathematicians of the day—it was surely known to the eminent Lambert—and, second, it was mentioned in two early histories of mathematics (Heilbronner, Leipzig, 1742 and Montucla, Paris, 1758), and elsewhere as well, before it fell into oblivion ([4], pages 36–39).

So the question ought to be generalized. If Saccheri himself did not put forth a claim to the discovery of non-Euclidean geometry (blinded, no doubt, by the ambition of reaching a different goal and—since he was overtaken by death in almost the very month that his book was published—deprived of an opportunity to re-evaluate his work), why did an entire generation of mathematicians, among them Lambert himself, not make the claim on Saccheri's behalf?

And then, Gauss. Nearly a century after Saccheri, Gauss (after years of effort) discovered non-Euclidean geometry. Although he fully understood the momentous nature of his discovery (and perhaps precisely because he understood it), he decided to say nothing about it, asked his friends to maintain silence, and made no public announcement of the discovery. (However, when Johann Bolyai's father communicated Johann's discoveries to Gauss, Gauss did not refrain from explaining that he himself had made the same discoveries earlier (see Gauss's letter to W. Bolyai ([2], page 100)).

Finally, Lobachevsky and Bolyai, whose discoveries were published in 1829–30 and 1832, respectively. The mathematical public paid little attention, and the work of both these men seemed destined to follow Saccheri's work into oblivion ([1], page 414). However, the posthumous publication (1860–1863) of Gauss's correspondence with Schumacher, and other factors, finally led to the public acceptance of non-Euclidean geometry—a good third of a century later.

We return to our original question. After deep thought and long effort having created non-Euclidean geometry, why did Saccheri not take the next, very simple step—that of recognizing the legitimacy of his creation? A recital of the above facts suggests that the inability to take that step was not peculiar to Saccheri himself; in his era, it

would seem, the existence of a valid geometry alternative to Euclid's was, quite literally, unthinkable. Not impossible, not wrong, but *unthinkable.*

Let us, then, compare Saccheri's era with ours. At the present day, we have an abundance of organized knowledge, which offers explanations—in which we have the fullest confidence—of many aspects of our physical universe. We need only refer to thermodynamics, geophysics, fluid dynamics, paleontology—the list is endless, and no one can possibly master all the knowledge that is available. But none of this knowledge extends back more than two hundred years. A group of educated eighteenth-century men, for example, sitting before an open fire, could no more understand or explain the nature of that fire than could their Neanderthal ancestors. For, the nature of light, the nature of heat, the nature of chemical combination and, in particular, of combustion, even the existence of oxygen, were yet to be discovered. Nor did Science, in the eighteenth century, at all inspire confidence, as it does today. Newton, for example, upon completing the second volume of his *Principia,* wrote to Edmund Halley: "The third I now design to suppress. Philosophy [i.e., Natural Philosophy] is such an impertinently litigious lady that a man had as good be engaged in lawsuits as have to do with her."

Let us therefore take a brief inventory of what existed in the eighteenth century to satisfy man's craving for certainty, for organized knowledge.

Physical science, as we have just mentioned, was not yet ready to satisfy this need for certainty. The teachings of the Church were indeed unquestioned Truths, for the Faithful. But the Faithful had to be aware that these Truths were ignored by much of mankind and were under constant attack by heretics. Philosophy seemed to offer certainty, but the existence of competing, and contradictory,

schools of philosophy betrayed an underlying uncertainty.

Contrast all of this with Geometry, which for two thousand years had been accepted as being the Science of the space in which we live. The precisely stated results of Geometry were proved logically on the basis of a small number of self-evident axioms of the utmost simplicity. Not only were elaborate geometrical relations proved but even quite simple ones—those so seemingly elementary as to hardly call for proof—were nonetheless fitted into a logical structure of proof. This elegant logical structure had been formulated by Euclid two thousand years earlier, and no one during those two thousand years had challenged it—not the ancient Greeks, nor the Arabs, nor the Europeans of the preceding centuries.

If a valid geometry, alternative to Euclid's were to exist, then Euclidean geometry would not necessarily be the science of space, and in fact there would no longer *be* a science of space. And with that science gone, there would be nothing—no science at all. Thus, in addition to the many reasons for not doubting Euclidean geometry to be the one and only geometry (after all, in our day, it is still the geometry of architecture, engineering, and most branches of science), we have another and powerful *silent* motive—a motive which does not reach consciousness and which for that reason is all the more powerful, the sort of motive which, under the right circumstances, makes an idea *unthinkable.*

This motive remained operative until the first modern science appeared on the scene—Celestial Mechanics. We recall the sequence of events: Newton explained the motion of the planets on the basis of his theory of gravitation. He was unable, however, to determine whether or not the planetary system was stable, and he freely admitted that the hand of God might have to intervene from time to time to set things right. It is to be noted that he

did *not* insist that in due course a proof of stability would be found. A century of mathematical research finally yielded a proof, and in the years 1799 to 1825 (1799, 1799, 1802, 1805, and 1823-25) Laplace initiated a new era in human thought by the publication of the five volumes of his epoch-making *Mécanique Céleste,* in which the new science was established.

Consider now the parallel events in the history of non-Euclidean geometry. Gauss made his discoveries between 1816 and 1831. Ferdinand Karl Schweikart handed Gauss a memorandum on what he called *Astral Geometry* (hyperbolic non-Euclidean geometry) in December of 1818. Schweikart's nephew, Taurinus, did valuable research on the subject between 1824 and 1826. Lobachevsky began his investigations after 1823 and published his discoveries in *Principles of Geometry* in 1829–30. On November 3, 1823, Johann Bolyai wrote to his father that he had made important discoveries, and his father urged immediate publication, writing that "it is right that no time be lost in making [them] public . . . first, because ideas pass easily from one to another . . . and, second, . . . many things have an epoch, in which they are found at the same time in several places, just as violets appear on every side in the Spring."

Let us now indicate how the Second Edition of *Euclides Vindicatus* differs from the First Edition. Only one or two minor changes have been made in the English text, and none in the Latin. Editor's Notes have been added on pages 245 ff. and the symbol ✲ has been inserted at those places in the English text to which the Editor's Notes make reference. The main content of the Notes are a series of annotations to Saccheri's text by the eminent scholars Paul Stäckel and Friedrich Engel. These annotations originally appeared as footnotes to these authors' German trans-

lation of *Euclides Vindicatus* ([4], pages 31–136) and were, with some modifications, translated into English by Professor F. Steinhardt. A second function of the notes is to make reference to Professor Halsted's own annotations, which appear here on pages 243–244. In addition, the Notes contain several alternative translations to a few sentences of the English text.

A.G.

BIBLIOGRAPHY

[1] KLINE, Morris. "New Geometries, New Worlds," *Mathematics in Western Culture.* New York: Oxford University Press, 1953, pp. 410–431.

[2] BONOLA, Roberto. *Non-Euclidean Geometry, A Critical and Historical Study of Its Development,* 2nd rev. ed. La Salle: Open Court Publishing Company, 1938.

[3] STRUIK, D.J. "Saccheri, (Giovanni) Girolamo," *Dictionary of Scientific Biography,* Vol. XII. New York: Charles Scribner's Sons, 1975.

[4] STÄCKEL, Paul and ENGEL, Friedrich. *Die Theorie der Parallellinien von Euklid bis auf Gauss.* Leipzig, 1895.

INTRODUCTION.

INTRODUCTION.

Giovanni Girolamo Saccheri was born at San Remo in the night between the 4th and 5th of September, 1667, as the authority on his life, the late Alberto Pascal tells us.

He was notably precocious.

March 24, 1685, he entered the Jesuit order. Toward 1690 he terminated the period of novitiate at Genoa and was sent by his superiors to the Collegio di Brera in Milan, to teach grammar and at the same time to study philosophy and theology. There the reading of the *Elements* of Euclid was recommended to him by the professor of mathematics, the Jesuit father Tommaso Ceva, from whose brother Giovanni the theorem Ceva is named.

In 1694 Saccheri was commanded to teach philosophy and polemic theology in the Collegio dei Gesuiti of Turin. In 1697 he was sent to Pavia. Says Pascal, "Fruit of these three years of philosophic teaching was a little book which well merits to be better known: perhaps its extreme rarity has contributed to this oblivion, even since Giovanni Vailati, in 1903, brought to light its superlative merit."

The few brief pages of Vailati, who died in 1909, furnish the only help on this *libretto,* an *opera dimenticata* first appearing with what he calls the *titolo abbastanza enigmatico*:

Logica demonstrativa, quam una cum Thesibus ex tota Philosophia decerptis, defendendam proponit Joannes Franciscus Casalette Graveriarum Comes sub auspiciis Regiae Celsitudinis Victorii Amedei II. Sabaudiae Ducis, Pedemontium Principis, Cypri Regis, etc.

Augustae Taurinorum Typis Joannis Baptistae Zappatae 1697. Superiorum permissu.

(In 16^{0}, pp. xii-287.)

Thus, in this first edition, the name of the author does not appear. Taking advantage of the examination of Count Gravere, then his student, Saccheri, with the theses, published his course in logic, letting it appear as if the count's.

The only existing copy is in the Biblioteca Nazionale of Milan (Colloc. B. X. 4854).

On it is written:

Auctore P're. Hyeronymo Saccherio Societatis Jesu,
and below:

Ex Biblioth'^{a} *Collegii Brayd'*^{is} *Soc'*^{is} *Jesu. Ins.*[criptus] *Catal'*^{o}.

Saccheri, astute and prudent, had his reasons for issuing this three-year child of his genius under the count's cloak. Then as professor he changed subjects and residence, and only four years afterward did the book appear with his name.

The first issue Saccheri never mentions. The second edition, so called, he refers to repeatedly and insistently. It differs from the first by some suppressions, especially in the preface, but no thought has been added during these four years of waiting.

Its title is:

Logica demonstrativa auctore Hieronymo Saccherio Societatis Jesu, olim in Collegio Taurinensi eiusdem Societatis Philosophiae, ac Theologiae Polemicae, nunc in Archigymnasio Ticinensi Publico Matheseos Professore.

Illustriss. Domino D. Philippo Archinto Sacr. Rom. Imp. Comiti, Marchioni Patronae, Comit. Trainati, Domino Erbae et Terrar. adiacen. Plebis Ticini, et Condom. Albizati, ac Reg. Duc. Senatori etc.

Ticini Regii. MDCCI.

Typis Haeredum Caroli Francisci Magrii Impressorum Civit. Superiorum permissu.

(In 8^{0}, pp. vi-167.)

At Cologne in 1735 appeared a third edition after the uthor's death and the publication of his *Euclides vindicatus* n 1733.

Its title is:

Logica demonstrativa, Theologicis, Philosophicis et Ma-hematicis Disciplinis accommodata; Auctore R. P. Hiero-ymo Saccherio, Societatis Jesu, olim in Collegio Taurin-nsi eiusdem Societatis Philosophiae ac Theologiae Pole-nicae; nunc in Archi-Gymnasio Ticinensi publico Mathe-eos professore.

Augustae Ubiorum, sumtu Henrici Noethen, Bibliopolae, n pladea vulgo dicta *unter Helmschläger* sub insigni *capitis urei.* MDCCXXXV.

(In 8°, pp. vi-162.)

The editor terminates a laudatory preface with the pigram:

> "Si tua, Saccheri, ingenio
> par penna fuisset,
> aetas ostendat vix tibi
> nostra parem."

In the Stadtbibliothek of Cologne (Augusta Ubiorum) the only existing copy of this posthumous edition.

In his preface our author says: "Quattuor in partes logi-am nostram, cum Aristotele, dividimus. Prima docebit egulas rectae argumentationis; secunda tradet methodum enendam in cognitionibus scientificis; tertia sternit viam d cognitiones opinativas; quarta fallacias detegit."

The scholastic logic undergoes a critical elaboration hich takes the form of a series of demonstrations based pon postulates and definitions and interconnected in a way nalogous to the method of geometers.[1]

In the same prelude mention is made of what he judges ew and important contributions to the ordinary treatment f logic.

[1] "Severa illa methodo quae primis principiis vix parcit nihilve on clarum, non evidens, non indubitatum, admittit.—Ea quam dixi ometriae severitas quae nihil indemonstratum recipiat." *Ibid.*

Says Heath: "Mill's account of the true distinction between *real* and *nominal* definitions had been fully anticipated by Saccheri."

In his *Logica demonstrativa* Saccheri lays down the clear distinction between what he calls *definitiones quid nominis* or *nominales,* and *definitiones quid rei* or *reales,* namely, that the former are only intended to explain the meaning that is to be attached to a given term, whereas the latter, besides declaring the meaning of a word, affirm at the same time the existence of the thing defined or, in geometry, the possibility of constructing it. The *definitio quid nominis* becomes a *definitio quid rei* "by means of a *postulate,* or when we come to the question whether the thing *exists* and it is answered affirmatively."[2]

Definitiones quid nominis are in themselves quite arbitrary, and neither require nor are capable of proof; they are merely provisional, and are only intended to be turned as quickly as possible into *definitiones quid rei,* either

1. by means of a postulate in which it is asserted or conceded that what is defined exists or can be constructed, e. g., in the case of *straight lines* and *circles,* to which Euclid's first three postulates refer, or

2. by means of a demonstration reducing the construction of the figure defined to the successive carrying-out of a certain number of those elementary constructions, the possibility of which is *postulated.* Thus *definitiones quid rei* are in general obtained as the result of a series of demonstrations.

Saccheri gives as an instance the construction of a square in Euclid I. 46.

Suppose that it is objected that Euclid had no right to define a square, as he does at the beginning of the Book, when it was not certain that such a figure exists; the objection, he says, could only have force if, before proving and making the construction, Euclid had assumed the afore-

[2] "Definitio *quid nominis* nata est evadere definitio *quid rei* per *postulatum* vel dum venitur ad quaestionem *an est* et respondetur affirmative." *Ibid.*

said figure as given. That Euclid is not guilty of this error is clear from the fact that he never presupposes the existence of the square as defined until after I. 46.

Confusion between the *nominal* and the *real* definition as thus described, i. e., the use of the former in demonstration before it has been turned into the latter by the necessary proof that the thing defined exists, is, according to Saccheri, one of the most fruitful sources of illusory demonstration, and the danger is greater in proportion to the "complexity" of the definition, i. e., the number of variety of the attributes belonging to the thing defined. For the greater is the possibility that there may be among the attributes some that are *incompatible,* i. e., the simultaneous presence of which in a given figure can be proved, by means of *other* postulates, etc., forming part of the basis of the science, to be impossible.

This signal anticipation of Mill's famous distinction would alone justify the only known protagonist of the *Logica demonstrativa* hitherto, Vailati, in saying of Saccheri: "This gives him the right to an eminent place in the history of modern logic."

But in additional elaboration Saccheri broadens the matter, clearly recognizing the more general question relative to the necessity of excluding the possible existence of incompatibility among the fundamental postulates made the basis of a demonstrative science; and not merely their directly contradicting one another, but whether the falsity of one of them could be proved by means of the others, a thing not directly recognizable.

These questions, far from having grown old, are acquiring an ever greater importance with the accentuation of the modern tendency to regard as the function of mathematics, the development, logically coherent, of the consequences flowing from a given system of premises, whether or no these be susceptible of a direct interpretation or experimental verification.

Since actually, in this case, the postulates assume the character of simple *hypotheses* subject only to the condition

of being mutually *compatible,* that is, of neither directly nor indirectly contradicting one another, the question relative to the means of ascertaining whether such compatibility really exists, ceases to be, in Vailati's phrase, a pure question *de luxe,* upon its solution having come to depend the legitimacy and even the possibility of assuming a given system of hypotheses as basis of a demonstrative science.

How high the merit of having been far the first to envisage this difficult matter and to have proffered an analysis of the various forms of fallacy to which its non-recognition may give rise!

Precisely to such subject is dedicated the final chapter of the *Logica demonstrativa.* And so ultra-modern and yet unfinished is this whole question here raised and entered upon first, that it beckons with rising interest to mathematicians and philosophers toward this little book so near to vanishing unrecognized from the earth.

"Huc usque de fallaciis communiter observatis. Duas adhuc superaddemus nec eas ut opinor parvi momenti. Hanc 'fallaciam complexi' appello, illam 'duplicis definitionis' seu 'hypothesis' " (p. 256 of 1st ed.).

The fallacy of "complex definitions," such as attribute to the thing defined the simultaneous possession of diverse properties, as for example Borelli's of "parallel" the property of being a straight line and that of being *also* the locus of points of a plane equidistant from another given straight, consists in supposing that such definitions can be adopted unchecked in the demonstrations, without the compatibility of the properties themselves having first been ascertained.

Obviously, in case such compatibility is lacking, in case the existence of an object possessing simultaneously the properties in question can be proved impossible (by means of the other hypotheses anteriorly postulated as basis of the demonstrative science under discussion), any argumentation among whose premises figure such definitions combined with the aforesaid hypotheses, ceases to have value being based on contradictory premises.

The forms of illusory reasoning examined and probed

y Saccheri under the name of *fallaciae duplicis hypothesis* re precisely those which consist in believing that consequences worth considering can be deduced from systems of ypotheses incompatible with one another (to wit, such hat among them are some whose negation can be deduced rom the others); and of such he passes in review various ypes, beginning with the simplest, namely, that of a syllogism whose premises directly contradict one another, and oing on to the more complicated cases in which the contradiction can be revealed only by the successive development f the consequences of the system of hypotheses, or postuates, assumed as basis of the entire matter.

In the investigation of the independence of postulates the nethod consists in finding a case or a particular interpreation in which the proposition we wish to prove not deduible from others given, ceases to be true while all the others emain true. If such is found, we conclude that the propoition *cannot be deduced* from these others, else it would be rue whenever they were.

This use and construction of examples to show the ndependence of a certain proposition from others given as of late assumed the importance of an ordinary and ndispensable procedure in the strictly rigorous elaboration f any deductive theory (in America, Robert L. Moore, Iuntington, Veblen, and others). But Saccheri was the rst to use this procedure, and constructs his example without leaving the field of formal logic. If his treatment of is "hypothesis of acute angle" is another case, it is the most narvelous in the world.

In his *Opus de proportionibus* (lib. V, prop. 201),[3] Cardan (born at Pavia, 1501), to prove that two sides of certain triangle he has occasion to consider are greater han a certain arc of a circle comprised between them, adopts peculiar reasoning and vaunts himself of this singular rocedure as an extraordinary discovery of his own:

[3] *Cardani Opera,* Lugduni [Lyons], 1663, t. IV, p. 579, published y Spon in ten volumes, folio.

"Hanc propositionem non scripsi, quod esset magni momenti, sed propter modum probandi.

"Si enim respicis, ex uno opposito (scilicet quod peripheria circuli sit major trianguli lateribus) ostendo, demonstratione non ducente ad inconveniens sed simplici, quod ipsa peripheria minor est trianguli lateribus.

"Et hoc nunquam fuit factum ab aliquo, immo videtur plane impossibile, et est res admirabilior quae inventa sit ab orbe condito, scilicet ostendere aliquod ex suo opposito demonstratione non ducente ad impossibile, et ita ut non possit demonstrari ea demonstratione nisi per illud suppositum quod est contrarium conclusioni, velut si quis demonstraret quod Socrates est albus quia est niger, et non possit demonstrare aliter; et ideo est longe majus Chrysippaeo Syllogismo."[4]

But that Cardan had been anticipated in this mode of deduction is twice noted by C. Clavius (1537-1612), once apropos of a demonstration given by Theodosius of Tripoli (*Sphaericorum,* lib. I. prop. 12), in proof of the theorem that two circles on the same sphere cannot bisect one another unless they are great circles:

"Hic vides mirabile argumentandi modus quod ex eo quod dicitur C non esse centrum sphaerae demonstratum est, demonstratione affirmativa, C esse centrum sphaerae. Quo modo argumentandi etiam usus est Euclides (IX. 12.) et Cardanus *De proportionibus* (V. prop. 201)";

and again in a scholium on Eu. IX. 12, in his *Euclidis elementorum libri XV*. Roma. 8°. 1574.

Euclid's proof is a characteristic example of this logical procedure, this type of demonstration.

[4] "And this has never been done by any one: nay, it seems clearly impossible, and is the most wonderful thing that has been devised since the creation of the world, namely, to prove something from its opposite, the demonstration not leading to an impossibility and in such a way that it could not be proved by that demonstration except by that being supposed which is contrary to the conclusion just as if one were to prove that Socrates is white because he is black and could not prove it in any other way; and for that reason it is far greater than the Chrysippæan Syllogism."

If there be how many numbers soever in continued proportion from unity: Then whatever prime numbers measure the last, the same will also measure that next after the unit.

Let there be as many numbers as we please, A, B, C, D continual proportionals from unity; I say whatever prime numbers measure D will measure A also.

	Unity	A	B	C	D
For example,	1	4	16	64	256
		E	H	G	F
		2	8	32	128

For let some prime number E measure D; I say E measures A.

For suppose it does not.

Now E is prime, and a prime number is prime to any it does not measure [VII. 29]; therefore E, A are prime to one another.

And since E measures D, let it measure it by the units in F; therefore E multiplying F produces D.

Again, since A measures D by the units in C [IX. 11 and Porism],[5] therefore A multiplying C produces D. But E has also by multiplying F made D; whence the product of A, C is equal to the product of E, F.

Therefore, as A to E, so is F to C [VII. 19].

But A, E are prime; primes are also least [VII. 21],[6] and the least measure those which have the same ratio the same number of times, the antecedent the antecedent and the consequent the consequent [VII. 20]; therefore E measures C.

Let it measure it by G; then E multiplying G produces C. But A has also by multiplying B made C [IX. 11 and Porism].

[5] $x^{m+n} = x^m x^n$.

[6] "Numbers prime to one another are the least of those which have the same ratio as they."

Therefore the product of A, B is equal to the produc of E, G.

Whence, as A to E, so is G to B [VII. 19].

But A, E are prime; primes are also least [VII. 21] and do equally measure those that have the same ratio a they, the antecedent the antecedent and the consequent th consequent [VII. 20]:

Wherefore E measures B.

Let it measure it by H; then E multiplying H produce B. But A has also by multiplying itself made B [IX. 8] therefore the product of E, H is equal to the product o A into itself.

Therefore, as E to A, so is A to H [VII. 19].

But A, E are prime; primes are also least [VII. 21], an the least measure those which have the same ratio the sam number of times, the antecedent the antecedent and th consequent the consequent [VII. 20]; therefore E measure A, as antecedent antecedent. Q. E. D.

The comment of Clavius is:

"Est autem res admirabilis huius propositionis demon stratio. Nam ex eo quod *b* dicatur non metiri ipsum *a* ostendit, demonstratione affirmativa, *b* ipsum *a* metiri, quo videtur fieri non posse. Nam si quis demonstrare institua Socratem esse album, ex eo quod non est albus, paradoxun aliquid et inopinatum in medio videatur afferre: cui tame non absimile quid factum hic est in numeris ab Euclide, e in aliis nonnullis propositionibus quae sequuntur."

Now the edition of Clavius was the Euclid recommende by T. Ceva to Saccheri.

This recondite legerdemain of logic, so striking to Carda and Clavius, seized with a more permanent fascination th adroit mind of the subtle Saccheri. It is the dominant not of the *Logica demonstrativa,* and thence adventures th quest of the holy grail, Euclid's Parallel Postulate.

This type of reasoning consists in assuming as hypothesis the falsity of the very proposition to be proved, and in show- ing how *also* when taking this hypothesis as point of de-

parture, none the less do we likewise arrive at the conclusion that the proposition in question is true.

In this process we reach nothing absurd or false, but thus proceeding we attain the very proposition which was to be proved, so that in this way it shows itself as a consequence of its own negation.[7]

In his preface, Saccheri, enumerating the parts of his work in which he believes himself to have made new and important contributions to the ordinary treatment of logic, gives first place to Chapter 11 of Part I, devoted precisely to this type of reasoning. He might still claim the merit of being not only the first but the only one to employ this method in a systematic treatise on logic. To have applied it to the elaboration of the rules of the scholastic logic, Saccheri regards as one of the most important ameliorations introduced by him in the treatment of the subject, and for us it is highly significant to note how the greatest advantage inherent in this innovation of his consists for him in his being able by this means to render his exposition independent of the assumption of a certain postulate he believes indispensable if the ordinary treatment be followed.

Hence there is an exact correspondence between the use he makes in his *Logica* of this demonstrative procedure, and the use he afterward attempted to make of it in his *Euclides vindicatus,* aiming to obviate the necessity of assuming the Parallel Postulate.

Vailati, of whose few precious pages we avail ourselves, points out that Saccheri tells how already in his youth he arrived at the idea that the characteristic property of the most fundamental propositions, in every demonstrative science, was precisely their being indemonstrable except by recourse to this IX—12 type of argumentation (see *Eu. vind.,* pp. 99-100), and then adds:

"It was in hopes of reaching in this way a proof of the Parallel Postulate, namely, deducing it from the very hypothesis of its falsity, that Saccheri pushed on in the in-

[7] "Sumam contradictoriam propositionum demonstrandarum ex eoque, ostensive ac directe, propositum eliciam." *Log. dem.,* p. 130.

vestigation of the consequences flowing from the other two alternative hypotheses to which the negation of the Parallel Postulate gave rise, attaining thus results fitting to carry on in their sequence to a discovery far more important than what he had in mind to reach, namely to the discovery of a wholly new geometry of which the old is only a simple particular case.

"In this regard, his position is not unworthy to be compared with that of his great fellow-countryman Columbus, who, precisely in hopes of being able to reach by a new way regions already known, was led to the discovery of a new continent."

Twisting Vailati's fine comparison, Saccheri's file of Indians turned out to be Columbians (Americans).

Realizing its importance as the coconut out of whose eyes the palm was to shoot up, which rises high above the flat and circumscribed old world, let us further look at the little *Logica*.

Its Chapter 9, of Part I, deduces the ordinary rules of the scholastic logic, relative to the conversion of propositions and to the construction of the various kinds of syllogisms, by a procedure imitating that followed by geometers in their treatises, and notes how in such procedure it is often necessary to have recourse to the assumption that, given any term, it is always possible to find other terms which are not coextensive with it nor with its negation.[8] It proposes then to re-elaborate the same subject following a method other and more refined (*aliam nobiliorem viam*) with which there is no need of using the mentioned postulate.

This aim is attained by taking as point of departure those among the propositions antecedently proved in whose demonstration no use has been made of the assumption in question, and by deducing from these, recourse being had to the IX—12 form of argument, the remaining propositions, which before had been obtained by using the assumption thus eliminated.

[8] "Postulatur non omnes terminos esse pertinentes mutua sequela aut repugnantia, sed quosdam esse inferiores et superiores, quosdam etiam impertinentes." *Log. dem.*, p. 30.

Here, for instance, is the demonstration *in nobiliorem viam* of the noted rule of the scholastic logic according to which in syllogisms of the so-called First Figure (in syllogisms, namely, in which the subject and the predicate of the conclusion enter respectively as subject and predicate also in the premises), the premise in which appears the subject of the conclusion cannot be negative: *In prima figura minor non potest esse negativa.*

Taking the simplest particular case, it is to be proved that from the two propositions:

Every A is a B,
No C is an A,

can be inferred no general proposition (affirmative or negative) having C as subject and B as predicate.

Proposing to demonstrate this rule by means of only the syllogism *Barbara,* which has its two premises universal and affirmative, Saccheri first observes that his aim would be attained if, for each of the different forms of syllogisms with negative minor premise constructible in the first figure, he could succeed in finding *examples* (that is, could choose such particular meanings for the terms entering) for which the two premises being true, the conclusion was false:

"Si quispiam syllogismus taliter constructus non recte concludit, nullus alius similiter constructus vi formae concludet" (*Logica dem.*, p. 130).

[Is this Italy, 1697, or America, 1919?]

For example, to prove that, from the two premises

Every A is a B,
No C is an A,

we cannot deduce the conclusion

No C is a B.

Attribute to the terms A, B, C, respectively the three following significations:

A = syllogism of the first figure, having the two premises universal and affirmative;

B = a valid syllogism;

C = syllogism of the first figure having one premise negative.

The two premises

Every A is a B,
No C is an A,

will then become the two following propositions:

1. Every syllogism of the first figure having the two premises universal and affirmative is a valid syllogism:
2. No syllogism of the first figure having one premise negative, is a syllogism of the first figure having the two premises universal and affirmative.

Now these two premises both being true, we must either admit as true the conclusion:

No C is a B

(that is, No syllogism of the first figure having one premise negative is a valid syllogism), or else concede that the meanings we have given to the terms A, B, C of the syllogism whose validity is in question, render true its two premises and false its conclusion.

In either case we are equally forced to admit that the syllogism in question is not valid.

"Vel concedis vel negas consequentiam. Si concedis, habetur intentum. Si negas, conclusioni dissentiens post concessas praemissas, fateris ipse legitimum non esse ex praemissis eiusmodi consequentiam quod intendebatur" (*Logica dem.*, p. 132).

His teaching of logic ended, his course published, its weapon of predilection, IX—12, left in his powerful hands, in his new field, mathematics, what heroic adventure was worthy its trenchant edge?

Sir Henry Savile, in his *Praelectiones tresdecim in principium Elementorum Euclidis habitae 1620,* Oxford, 4°, 1621, p. 140, says: "In pulcherrimo Geometriae corpore duo sunt naevi."

And the greatest of these moles is the eternal Parallel

Postulate. Here then is something worthy of Saccheri's steel. To prove it from its own denial!

This would show that Euclid's assumptions, though compatible, were not all independent. On the other hand, the independence of the Parallel Postulate from the other assumptions would be established if it were shown to be indemonstrable from them even with the help of its own contradictory opposite, that is, even by means of Saccheri's darling type, IX—12.

To get this negative in convenient form, Saccheri uses for it an equivalent, employing a figure found in his Clavius, 1574, and again in Giordano Vitale da Bitonto, in his *Euclide restituto overo gli antichi elementi geometrici ristaurati, e facilitati. Libri XV*. Roma. fol., 1680; and in both works precisely in discussion of this very matter. The figure is the isosceles bi-rectangular quadrilateral. Its other two angles are equal. To assume one right is, with Euclid's other assumptions, equivalent to the parallel postulate, whose negative therefore is to assume one oblique. Armed then with this form of its denial as an addition to the other Euclidean assumptions, Saccheri sallies forth to the fray, steeled for victory or defeat—but not for the wholly unexpected and to him inexplicable compound of victory and defeat which he met.

His negation breaks into two equal parts. The angle assumed oblique is either obtuse or acute. If it be obtuse, he easily achieves his accustomed victory: he proves the Parallel Postulate. If it be acute, this twin will not win.

Why? We know. Saccheri never did.

Besides the Archimedes assumption, Euclid, and every one else for more than a century after Saccheri, assumes that the straight line is of infinite length. These assumptions nullify the possibility of a pair of obtuse angles in a bi-rectangular isosceles quadrilateral, and to that extent prove the "hypothesis of right angle," which is then equivalent to the Parallel Postulate. But they are no obstacle to this pair of angles being acute.

Had there been some other unconscious assumption of

Euclid's, preventing their being acute, then Saccheri might well have declared the Parallel Postulate completely demonstrated.

But there is none.

Under the "hypothesis of acute angle" the chain of beautiful theorems developed, grew, but did not end.

So flowered the beauteous body of a new geometry, mermaid-like, the latter portions somewhat fishy, but oh! the elegant torso.

Of this book says the genius Corrado Segre: "Nevertheless the first seventy pages (apart from a few isolated phrases), up to Proposition 32 inclusive, constitute an ensemble of logic and of geometric acumen which may be called *perfect.*"

EUCLIDES VINDICATUS

EUCLIDES
AB OMNI NÆVO VINDICATUS:
SIVE
CONATUS GEOMETRICUS
QUO STABILIUNTUR
Prima ipsa universæ Geometriæ Principia.

AUCTORE

HIERONYMO SACCHERIO
SOCIETATIS JESU
In Ticinensi Universitate Matheseos Professore.

OPUSCULUM

EX.^{MO} SENATUI
MEDIOLANENSI
Ab Auctore Dicatum.

MEDIOLANI, MDCCXXXIII.

Ex Typographia Pauli Antonii Montani. *Superiorum permissu*

EUCLID

FREED OF ALL BLEMISH

OR

A GEOMETRIC ENDEAVOR IN WHICH ARE ESTABLISHED THE FOUNDATION PRINCIPLES OF UNIVERSAL GEOMETRY

BY

GIROLAMO SACCHERI

OF THE SOCIETY OF JESUS

PROFESSOR OF MATHEMATICS IN THE UNIVERSITY OF PAVIA.

A WORK DEDICATED TO
THE NOBLE SENATE OF
MILAN BY THE AUTHOR

MILAN, 1733

PAOLO ANTONIO MONTANO SUPERIORUM PERMISSU

PROŒMIUM AD LECTOREM.

Quanta sit Elementorum Euclidis praestantia, ac dignitas, nemo omnium, qui Mathematicas disciplinas noverint, ignorare potest. Lectissimos hanc in rem testes adhibeo Archimedem, Apollonium, Theodosium, aliosque pene innumeros, ad haec usque nostra tempora rerum Mathematicarum Scriptores, qui non aliter haec Euclidis Elementa usurpant, nisi ut principia jam diu stabilita, ac penitus inconcussa. Verum tanta haec nominis celebritas vetare non potuit, quin multi ex Antiquis pariter, ac Recentioribus, iique Magni Geometrae naevos quosdam a se deprehensos censuerint in his ipsis pulcherrimis, nec unquam satis laudatis Elementis. Tres autem hujusmodi naevos designant, quos statim subnecto.

Primus respicit definitionem parallelarum, et sub ea Axioma, quod apud Clavium est decimumtertium Libri primi, ubi Euclides sic pronunciat: *Si in duas rectas lineas, in eodem plano existentes recta incidens linea duos ad easdem partes internos angulos minores duobus rectis cum eisdem efficiat, duae illae rectae lineae ad eas partes in infinitum protractae inter se mutuo incident.* Porro nemo est, qui dubitet de veritate expositi Pronunciati; sed in eo unice Euclidem accusant, quod nomine Axiomatis usus fuerit, quasi nempe ex solis terminis rite perspectis sibi ipsi faceret fidem. Inde autem non pauci (retenta caeteroquin Euclidaea parallelarum definitione)

PREFACE TO THE READER. ✡[1]

✡ Of all who have learned mathematics, none can fail to know how great is the excellence and worth of Euclid's *Elements.* As erudite witnesses here I summon Archimedes, Apollonius, Theodosius, and others almost innumerable, writers on mathematics even to our times, who use Euclid's *Elements* as foundation long established and wholly unshaken. But this so great celebrity has not prevented many, ancients as well as moderns, and among them distinguished geometers, maintaining they had found certain blemishes in these most beauteous nor ever sufficiently praised *Elements.* Three such flecks they designate, which now I name.

The first pertains to the definition of parallels and with it the axiom which in Clavius is the thirteenth of the First Book, where Euclid says:

If a straight line falling on two straight lines, lying in the same plane, make with them two internal angles toward the same parts less than two right angles, these two straight lines infinitely produced toward those parts will meet each other.

No one doubts the truth of this proposition; but solely they accuse Euclid as to it, because he has used for it the name axiom, as if obviously from the right understanding of its terms alone came conviction. Whence not a few (withal retaining Euclid's definition of parallels) have

[1]This symbol refers the reader to the Editor's Notes, on page 245 ff.

illius demonstrationem aggressi sunt ex iis solis Propositionibus Libri primi Euclidaei, quae praecedunt vigesimam nonam, ad quam scilicet usui esse incipit controversum Pronunciatum. [x]

Sed rursum; quoniam antiquorum in hanc rem conatus visi non sunt adamussim scopum attingere; factum idcirco est, ut multi proximiorum temporum eximii Geometrae, idem pensum aggressi, necessariam censuerint novam quandam parallelarum definitionem. Itaque; cum Euclides parallelas rectas lineas definiat, *quae in eodem plano existentes, si in utranque partem in infinitum producantur, nunquam inter se mutuo incidunt;* postremis expositae definitionis vocibus has alias substituunt: *Semper inter se aequidistant*; adeo ut nempe singulae perpendiculares ab uno quolibet unius illarum puncto ad alteram demissae aequales inter se sint.

At nova rursum hinc oritur scissura. Nam aliqui, et ii sane acutiores, demonstrare conantur parallelas rectas lineas prout sic definitas, unde utique gradum faciant ad demonstrandum sub ipsis Euclidaeis vocibus controversum Pronunciatum, cui nimirum ab ea vigesima nona Libri primi Euclidaei (pauculis quibusdam exceptis) universa innititur Geometria. Alii vero (non sine magno in rigidam Logicam peccato) eas tales rectas lineas parallelas, nimirum *aequidistantes,* assumunt tanquam datas, ut inde gradum faciant ad reliqua demonstranda.

Et haec quidem satis sunt ad praemonendum Lectorem super iis, quae materiam exhibebunt Libro priori hujus mei Opusculi: Nam uberior praedictorum omnium explicatio habebitur in Scholiis post Prop. vigesimam primam enunciati Libri, quem dividam in duas veluti partes. In priore imitabor antiquiores illos Geometras, nihil propterea sollicitus de natura, aut nomine illius lineae, quae omnibus suis punctis a quadam supposita recta linea aequidistet: Sed unice in id incumbam, ut con-

attempted its demonstration from those propositions of Euclid's First Book alone which precede the twenty-ninth, wherein begins the use of the controverted proposition. [x]

But again, since the endeavors of the ancients in this matter do not seem to attain the goal, so it has happened that many distinguished geometers of ensuing times, attacking the same idea, have thought necessary a new definition of parallels. Thus, while Euclid defines parallels as straight lines *lying in the same plane, which, if infinitely produced toward both sides, nowhere meet,* they substitute for the last words of the given definition these others: *always equidistant from each other*; so that all perpendiculars from any points on one of them let fall upon the other are equal to one another.

But again here arises a new fissure. For some, and these surely the keenest, endeavor to demonstrate the existence of parallel straight lines as so defined, whence they go up to the proof of the debated proposition as stated in Euclid's terms, upon which truly from that twenty-ninth of Euclid's First Book (with some very few exceptions) all geometry rests. But others (not without gross sin against rigorous logic) assume such parallel straight lines, forsooth *equidistant,* as if given, that thence they may go up to what remains to be proved.

And this is enough to indicate to the reader what will be the material of the First Book of this work of mine: for a more complete explication of all that has been said will be given in the scholia after the twenty-first proposition of this Book.

I divide this Book into two parts. In the First Part I will imitate the antique geometers, and not trouble myself about the nature or the name of that line which at all its points is equidistant from a certain line supposed straight; but merely undertake without any *petitio prin-*

troversum Euclidaeum Axioma citra omnem petitionem principii clare demonstrem; nunquam idcirco adhibens ex ipsis prioribus Libri primi Euclidaei Propositionibus, non modo vigesimam septimam, aut vigesimam octavam, sed nec ipsas quidem decimam sextam, aut decimam septimam, nisi ubi clare agatur de triangulo omni [xi] ex parte circumscripto. Tum in posteriore parte, ad novam ejusdem Axiomatis confirmationem demonstrabo non nisi rectam lineam esse posse, quae omnibus suis punctis a quadam supposita recta linea aequidistet. Horum autem occasione prima ipsa universae Geometriae Principia rigido examini subjicienda hic esse nullus est, qui non videat.

Transeo ad alios duos naevos Euclidi objectos. Prior respicit definitionem sextam Libri quinti super aeque proportionalibus: Posterior Definitionem quintam Libri sexti super compositione rationum. Hic autem erit secundi mei Libri unicus scopus, ut dilucide explicem praefatas Euclidaeas Definitiones, simulque ostendam non aequo jure hac in parte Euclidis nomen vexatum fuisse.

Prodest tamen rursum praemonere, demonstratum a me iri hac occasione unum quoddam Axioma, quod tutissime per omnem Geometriam versetur, sine indigentia illius *Postulati,* sub nomine Axiomatis ab interpretibus (ut reor) intrusi, cujus usus incipit ad 18. quinti. [xii]

cipii clearly to demonstrate the disputed Euclidean axiom. Therefore never will I use from those prior propositions of Euclid's First Book, not merely the twenty-seventh or the twenty-eighth, but not even the sixteenth or the seventeenth, except where clearly it is question of a triangle every [xi] way restricted.

Then in the Second Part for a new confirmation of the same axiom, I shall demonstrate that the line which at all its points is equidistant from an assumed straight line can only be a straight line. But every one sees that on this occasion the very first principles of all geometry are to be subjected to a rigid examination.

I go on to the other two blemishes charged against Euclid. The first pertains to the sixth definition of the Fifth Book about proportionals; the second to the fifth definition of the Sixth Book about the composition of ratios. It will be the sole aim of my Second Book to clearly expound the Euclidean definitions mentioned, and at the same time to show that Euclid's fame is here unjustly attacked.

Yet again it is well to state that on this occasion I shall prove a certain axiom that may safely be applied throughout the whole of geometry, without need of that *postulate,* put in (as I believe) by commentators under the name of axiom, whose use begins at the eighteenth proposition of the Fifth Book. [xii]

INDICIS LOCO

ADDENDA CENSEO, QUAE SEQUUNTUR.

1. In I. et II. Propos. Lib. primi duo jaciuntur principia, ex quibus in III. et IV. demonstratur, angulos interiores ad rectam jungentem extremitates aequalium perpendiculorum, quae ex duobus punctis alterius rectae, veluti basis, versus easdem partes (in eodem plano) erigantur, non modo fore inter se aequales, sed praeterea aut rectos, aut obtusos, aut acutos, prout illa jungens aequalis fuerit, aut minor, aut major praedicta basi: Atque ita vicissim. *a pag.* 1

2. Hinc sumitur occasio secernendi tres diversas hypotheses, unam anguli recti, alteram obtusi, tertiam acuti: circa quas in V. VI. et VII. demonstratur, unam quamlibet harum hypothesium fore semper unice veram, si nimirum depraehendatur vera in uno quolibet casu particulari. *a pag.* 5

3. Tum vero; post interpositas tres alias necessarias Propositiones; demonstratur in XI. XII. ac XIII. universalis veritas noti Axiomatis, respectu habito ad priores duas hypotheses, unam anguli recti, et alteram obtusi; ac tandem in XIV. ostenditur absoluta falsitas hypothesis

IN PLACE OF AN INDEX

SHOULD BE ADDED, I THINK, WHAT FOLLOWS.

1. In Propp. I. and II. of the First Book two principles are established, from which in Propp. III. and IV. is proved, that interior angles at the straight joining the extremities of equal perpendiculars erected toward the same parts (in the same plane) from two points of another straight, as base, not merely are equal to each other, but besides are either right or obtuse or acute according as that join is equal to, or less, or greater than the aforesaid base: and inversely. *From page* 1 *on.*

2. Hence occasion is taken to distinguish three different hypotheses, one of right angle, another of obtuse, a third of acute: about which in Propp. V., VI., and VII. is proved, that any one of these hypotheses is always alone true if it is found true in any one particular case. *From page* 5 *on.*

3. Then after the interposition of three other necessary propositions, is proved in Propp. XI., XII., and XIII., the universal truth of the famous axiom, respect being had to the first two hypotheses, one of right angle, and the other of obtuse; and at length in P. XIV. is shown the absolute falsity of the hypothesis of obtuse

anguli obtusi. Atque hinc incipit diuturnum proelium adversus hypothesin anguli acuti, quae sola renuit veritatem illius Axiomatis. [xiii] *a pag.* 10

4. Itaque in XV. ac XVI. demonstratur stabilitum iri hypotheses aut anguli recti, aut obtusi, aut acuti, ex quolibet triangulo rectilineo, cujus tres simul anguli aequales sint, aut majores, aut minores duobus rectis; ac similiter ex quolibet quadrilatero rectilineo, cujus quatuor simul anguli aequales sint, aut majores, aut minores quatuor rectis. *a pag.* 20

5. Sequuntur quinque aliae Propositiones, in quibus demonstrantur alia indicia pro secernenda vera hypothesi a falsis. *a pag.* 23

6. Accedunt quatuor principalia Scholia; in quorum postremo exhibetur figura quaedam geometrica, ad quam fortasse respexit Euclides, ut suum illud Pronunciatum assumeret tanquam per se notum. In tribus prioribus ostenditur non valuisse ad intentum praecedentes insignium Geometrarum conatus. Sed quia controversum Axioma exactissime demonstratur ex duabus praesuppositis rectis lineis *aequidistantibus;* monet ibi Auctor contineri in eo praesupposito manifestam petitionem *Principii.* Quod si provocari hic velit ad communem persuasionem, atque item exploratissimam *praxim;* rursum monet provocari non debere ad experientiam, quae respiciat puncta infinita, cum satis esse possit unica experientia uni cuivis puncto affixa. Quo loco tres ab ipso afferuntur invictissimae Demonstrationes Physico-Geometricae. *a pag.* 29

7. Supersunt duodecim aliae Proposi- [xiv] tiones, quae primae Parti hujus Libri finem imponunt. Non expono particularia assumpta, quia

angle. And here begins a lengthy battle against the hypothesis of acute angle, which alone opposes the truth of that axiom. [xiii] *From page* 10 *on.*

4. And so in Propp. XV. and XVI. is proved that the hypothesis either of right angle, or obtuse, or acute is established from any rectilineal triangle whose three angles together are equal to, or greater, or less than two right angles; and in like way from any rectilineal quadrilateral, whose four angles are together equal to, or greater, or less than four right angles. *From page* 20 *on.*

5. Five other propositions follow, in which are proved other indications for distinguishing the true hypothesis from the false. *From page* 23 *on*

6. Now come four fundamental scholia. In the last is exhibited a certain geometric figure, of which Euclid perhaps thought, in order that his proposition might assume self-evidence. In the preceding three is shown that the prior endeavors of distinguished geometers have not reached their aim. Since however the debated axiom can be exactly proved from two straight lines presupposed *equidistant,* the author here shows a manifest *petitio principii* to be contained in that presupposition. If one wishes here to appeal to common persuasion, and surest *experience,* again he shows appeal should not be taken to an experience involving an infinity of points, when a single experiment pertaining to any one point can suffice. In this place are set forth by him three invincible physico-geometric demonstrations. *From page* 29 *on.*

7. To the end of the First Part of this Book there remain twelve other propositions. [xiv] I do not state the particular assumptions, be-

nimis implexa. Solum dico ibi tandem manifestae falsitatis redargui inimicam hypothesim anguli acuti, utpote quae duas rectas agnoscere deberet, quae in uno eodemque puncto commune recipe-rent in eodem plano perpendiculum: Quod qui-dem naturae lineae rectae repugnans esse de-monstratur per quinque Lemmata, in quibus concluduntur quinque principalia Geometriae Axiomata, quae respiciunt lineam rectam, ac cir-culum, cum suis correlativis Postulatis. *a pag.* 43

8. Secunda pars continet sex Propositiones. Ibi autem; post expensam (juxta hypothesim anguli acuti) naturam illius lineae, quae omni-bus suis punctis a quadam praesupposita recta linea aequidistet; multis modis ostenditur, eam fore aequalem contrapositae basi, unde infertur praenunciatae hypothesis certissima falsitas. Quare tandem in ultima Propos. quae est XXXIX. exactissime demonstratur celebre illud Euclidaeum Axioma, cui nempe (ut omnes sci-unt) universa Geometria innititur. *a pag.* 87

9. Secundus Liber digeri commode non potuit in Propositiones, etiamsi locis opportunis plura intermista sint utilissima Theoremata, ac Problemata. Meretur nihilominus expresse no-tari unum quoddam Axioma, cujus ibi demon-stratur non modo veritas, verum etiam univer-salis utilitas [xv] pro omni Geometria, sine indi-gentia alterius parum decori Postulati, quod ab interpretibus censeri potest intrusum sub nomine Axiomatis, cujus nempe usus incipit ad 18. quinti. Et id quidem pro prima Parte hujus Libri, in qua vindicatur Def. sexta quinti Euclidaei. *a pag.* 102

cause they are too complex. I only say here at length I have convicted the hostile hypothesis of acute angle of manifest falsity, since it must lead to the recognition of two straight lines which at one and the same point have in the same plane a common perpendicular. That this is contrary to the nature of the straight line is proved by five lemmas, in which are contained five fundamental axioms relating to the straight line and circle, with their correlative postullates. *From page* 43 *on.*

8. The Second Part contains six propositions. Here, after investigating the nature (assuming the hypothesis of acute angle) of that line which at all its points is equidistant from an assumed straight line, it is shown in many ways that it equals the base opposite, whence is inferred the certain falsity of the aforesaid hypothesis. Wherefore at length in the last proposition, P. XXXIX., is exactly proved that famous axiom of Euclid, upon which (as everybody knows) the whole of geometry rests. *From page* 87 *on.*

9. The Second Book cannot conveniently be divided into propositions, although at opportune places are intercalated many most useful theorems and problems. Nevertheless is worthy of express mention a certain axiom, of which not merely the truth is there demonstrated but also the universal utility for all geometry, without need of the other inelegant postulate supposably inserted by commentators under the name of axiom, whose use begins at Eu. V. 18. So much for the First Part of this Book, in which is defended Eu. V. def. 6. *From page* 102 *on.*

10. Tum in secunda Parte; praeter nonnulla alia opportune addita, ad tuendas reliquas Definitiones ejusdem Quinti circa magnitudines proportionales; demonstratur priore loco (respectu habito ad magnitudines commensurabiles) quinta Definitio Sexti, etiamsi recipi ea deberet in *quid rei,* veluti Axioma: Sed rursum multis exemplis, ex ipso Euclide petitis, ostenditur nullius demonstrationis indigam eam esse, quia Definitionem *puri nominis.* Atque ita, post opportunam additam Appendicem, huic Operi finis imponitur. *a pag.* 132

10. Then in the Second Part, besides some other things opportunely added regarding other definitions of Eu. V about proportional magnitudes, is demonstrated in the first place (with respect to commensurable magnitudes) Eu. VI. def. 5, even if it ought to be taken in *quid rei* like an axiom. But on the contrary is shown by many examples drawn from Euclid himself that this needs no demonstration, because a definition *puri nominis*. And so after an Appendix opportunely added, an end is put to this work.

From page 132 *on* [*to page* 142].

EUCLIDIS AB OMNI NAEVO VINDICATI

LIBER PRIMUS:

IN QUO DEMONSTRATUR: DUAS QUASLIBET IN EODEM PLANO EXISTENTES RECTAS LINEAS, IN QUAS RECTA INQUAEPIAM INCIDENS DUOS AD EASDEM PARTES INTERNOS ANGULOS EFFICIAT DUOBUS RECTIS MINORES, AD EAS PARTES ALIQUANDO INVICEM COITURAS, SI IN INFINITUM PRODUCANTUR.

PARS PRIMA

PROPOSITIO I.

Si duae aequales rectae (fig. 1.) AC, BD, aequales ad easdem partes efficiant angulos cum recta AB: Dico angulos ad junctam CD aequales invicem fore.

Demonstratur. Jungantur AD, CB. Tum considerentur triangula CAB, DBA. Constat (ex quarta primi) aequales fore bases CB, AD. Deinde considerentur triangula ACD, BDC. Constat (ex octava primi) aequales fore angulos ACD, BDC. Quod erat demonstrandum.

EUCLID FREED OF ALL BLEMISH

BOOK I.

IN WHICH IS PROVED: ANY TWO COPLANAR STRAIGHT LINES, FALLING UPON WHICH ANY STRAIGHT MAKES TOWARD THE SAME PARTS TWO INTERNAL ANGLES LESS THAN TWO RIGHT ANGLES, AT LENGTH MEET EACH OTHER TOWARD THOSE PARTS, IF INFINITELY PRODUCED.

PART I.

PROPOSITION I.

If two equal straights [sects] (*fig.* 1) *AC, BD, make with the straight AB angles equal toward the same parts: I say that the angles at the join CD will be mutually equal.*

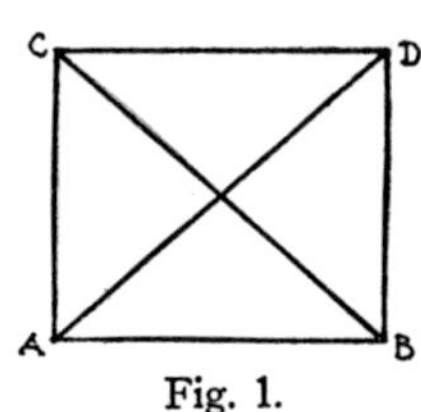

Fig. 1.

PROOF. Join AD, CB. Then consider the triangles CAB, DBA. It follows (Eu. I. 4) that the bases CB, AD will be equal.

Then consider the triangles ACD, BDC. It follows (Eu. I. 8) that the angles ACD, BDC will be equal.

Quod erat demonstrandum.

PROPOSITIO II.

Manente uniformi quadrilatero ABCD, latera AB, CD, bifariam dividantur (fig. 2.) in punctis M, et H. Di- [2] *co angulos ad junctam MH fore hinc inde rectos.*

Demonstratur. Jungantur AH, BH, atque item CM, DM. Quoniam in eo quadrilatero anguli A, et B positi sunt aequales, atque item (ex praecedente) aequales sunt anguli C, et D; constat ex quarta primi (cum alias nota sit aequalitas laterum) aequales fore in triangulis CAM, DBM, bases CM, DM; atque item, in triangulis ACH, BDH, bases AH, BH. Quare; collatis inter se triangulis CHM, DHM, ac rursum inter se triangulis AMH, BMH; constabit (ex octava primi) aequales invicem fore, atque ideo rectos angulos hinc inde ad puncta M, et H. Quod erat demonstrandum.

PROPOSITIO III.

Si duae aequales rectae (fig. 3.) AC, BD, perpendiculariter insistant cuivis rectae AB: Dico junctam CD aequalem fore, aut minorem, aut majorem ipsa AB, prout anguli ad eandem CD, fuerint aut recti, aut obtusi, aut acuti.

Demonstratur prima pars. Existente recto utroque angulo C, et D; sit, si fieri potest, alterutra ipsarum, ut DC, major altera BA. Sumatur in DC portio DK aequa-

PROPOSITION II.

Retaining the uniform quadrilateral ABCD, bisect the sides AB, CD (fig. 2) in the points M and H. [2] *I say the angles at the join MH will then be right.*

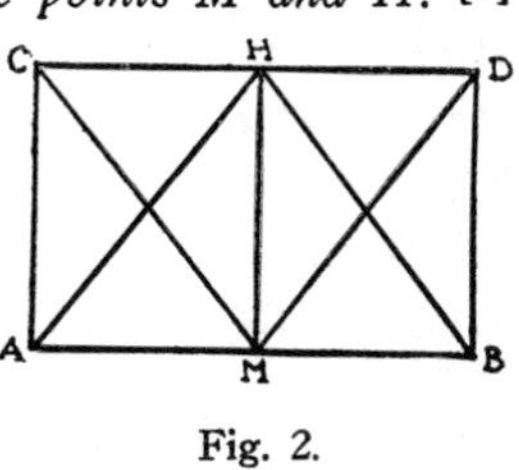

Fig. 2.

PROOF. Join AH, BH, and likewise CM, DM.

Because in this quadrilateral the angles A and B are taken equal and likewise (from the preceding proposition) the angles C, and D are equal; it follows (Eu. I. 4) (noting the equality of the sides) that in the triangles CAM, DBM, the bases CM, DM will be equal; and likewise, in the triangles ACH, BDH, the bases AH, BH.

Therefore; comparing the triangles CHM, DHM, and in turn the triangles AMH, BMH; it follows (Eu. I. 8) that we have mutually equal, and therefore right, the angles at the points M, and H.

Quod erat demonstrandum.

PROPOSITION III. ✡

If two equal straights [sects] *(fig. 3) AC, BD, stand perpendicular to any straight AB: I say the join CD will be equal to, or less, or greater than AB, according as the angles at CD are right, or obtuse, or acute.*

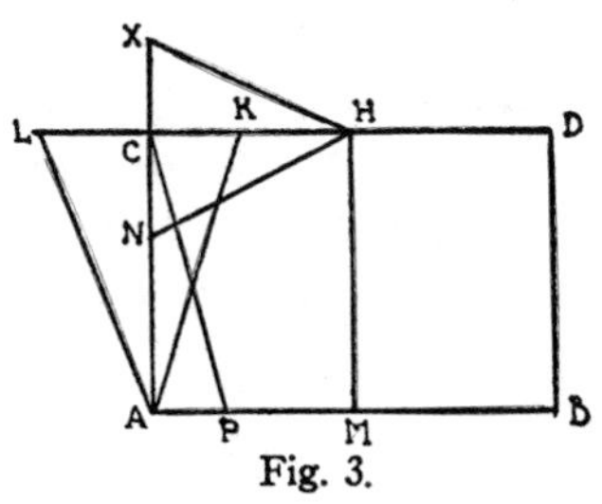

Fig. 3.

PROOF OF THE FIRST PART. Each angle C, and D, being right; suppose, if it were possible, either one of those, as DC, greater than the other BA.

lis ipsi BA, jungaturque AK. Quoniam igitur super BD perpendiculariter insistunt aequales rectae BA, DK, aequales erunt (ex prima hujus) anguli BAK, DKA. Hoc autem absurdum est; cum angulus BAK sit ex constructione minor supposito recto BAC; et angulus DKA sit ex constructione externus, atque ideo (ex decimasexta primi) major interno, et opposito DCA, qui supponitur rectus. Non ergo alterutra praedictarum rectarum, DC, BA, est altera major, dum anguli ad junctam CD sint recti; ac propterea aequales invicem sunt. Quod erat primo loco demonstrandum. [3]

Demonstratur secunda pars. Si autem obtusi fuerint anguli ad junctam CD, dividantur bifariam AB, et CD in punctis M, et H, jungaturque MH. Quoniam ergo super recta MH perpendiculariter insistunt (ex praecedente) duae rectae AM, CH, poniturque ad junctam AC angulus rectus in A, non erit (ex prima hujus) recta CH aequalis ipsi AM, cum desit angulus rectus in C. Sed neque erit major: caeterum sumpta in HC portione KH aequali ipsi AM, aequales forent (ex prima hujus) anguli ad junctam AK. Hoc autem absurdum est, ut supra. Nam angulus MAK est minor recto; et angulus HKA est (ex decimasexta primi) major obtuso, qualis supponitur internus, et oppositus HCA. Restat igitur, ut CH, dum anguli ad junctam CD ponantur obtusi, minor sit ipsa AM; ac propterea prioris dupla CD minor sit posterioris dupla AB. Quod erat secundo loco demonstrandum.

Demonstratur tertia pars. Tandem vero, si acuti fuerint anguli ad junctam CD, ducta pariformiter (ex praecedente) perpendiculari MH, sic proceditur. Quoniam

Take in DC the piece DK equal to BA, and join AK. Since therefore on BD stand perpendicular the equal straights BA, DK, the angles BAK, DKA will be equal (P. I.). But this is absurd; since the angle BAK is by construction less than the assumed right angle BAC; and the angle DKA is by construction external, and therefore (Eu. I. 16) greater than the internal and opposite DCA, which is supposed right. Therefore neither of the aforesaid straights, DC, BA, is greater than the other, whilst the angles at the join CD are right; and therefore they are mutually equal.

Quod erat primo loco demonstrandum. [3]

PROOF OF THE SECOND PART. But if the angles at the join CD are obtuse, bisect AB, and CD, in the points M, and H, and join MH.

Since therefore on the straight MH stand perpendicular (P. II.) the two straights AM, CH, and at the join AC is a right angle at A, the straight CH will not be (P. I.) equal to this AM, since a right angle is lacking at C. ✲

But neither will it be greater: otherwise in HC the piece KH being assumed equal to this AM, the angles at the join AK will be (P. I.) equal.

But this is absurd, as above. For the angle MAK is less than a right; and the angle HKA is (Eu. I. 16) greater than an obtuse, such as the internal and opposite HCA is supposed.

It remains therefore, that CH, whilst the angles at the join CD are taken obtuse, is less than this AM; and therefore CD double the former is less than AB double the latter.

Quod erat secundo loco demonstrandum.

PROOF OF THE THIRD PART. Finally, however, if the angles at the join are acute, MH being constructed as before perpendicular (P. II.), we proceed thus. Since on

super recta MH perpendiculariter insistunt duae rectae AM, CH, ponitur que ad junctam AC angulus rectus in A, non erit (ut supra) recta CH aequalis ipsi AM, cum desit angulus rectus in C. Sed neque erit minor: caeterum; si in HC protracta sumatur HL aequalis ipsi AM; aequales forent (ut supra) anguli ad junctam AL. Hoc autem absurdum est. Nam angulus MAL est ex constructione major supposito recto MAC; et angulus HLA est ex constructione internus, et oppositus, atque ideo minor (ex decimasexta primi) externo HCA, qui supponitur acutus. Restat igitur, ut CH, dum anguli ad junctam CD sint acuti, major sit ipsa AM, atque ideo prioris dupla CD major sit posterioris dupla AB. Quod erat tertio loco demonstandum.

Itaque constat junctam CD aequalem fore, aut mino- [4] rem, aut majorem ipsa AB, prout anguli ad eandem CD fuerint aut recti, aut obtusi, aut acuti. Quae erant demonstranda.

COROLLARIUM I.

Hinc in omni quadrilatero continente tres quidem angulos rectos, et unum obtusum, aut acutum, latera adjacentia illi angulo non recto minora sunt, alterum altero, lateribus contrapositis, si ille angulus sit obtusus, majora autem, si sit acutus. Id enim demonstratum jam est de latere CH relate ad contrapositum latus AM; similique modo ostenditur de latere AC relate ad contrapositum latus MH. Cum enim rectae AC, MH, perpendiculares sint ipsi AM, nequeunt (ex prima hujus) esse invicem aequales, propter inaequales angulos ad junctam CH. Sed neque (in hypothesi anguli obtusi in C) potest quae-

the straight MH stand perpendicular two straights AM, CH, and at the join AC is a right angle at A, the straight CH will not be equal to this AM (as above), since the angle at C is not right.✡ But neither will it be less: otherwise, if in HC produced HL is taken equal to this AM, the angles at the join AL will be (as above) equal.

But this is absurd. For the angle MAL is by construction greater than the assumed right MAC; and the angle HLA is by construction internal, and opposite, and therefore less than (Eu. I. 16) the external HCA, which is assumed acute.

It remains therefore, that CH, whilst the angles at the join CD are acute, is greater than this AM, and therefore CD the double of the former is greater than AB the double of the latter.

Quod erat tertio loco demonstrandum.

Therefore it is established that the join CD will be equal to, or less, [4] or greater than this AB, according as the angles at the same CD are right, or obtuse, or acute.

Quae erant demonstranda.

COROLLARY I.

Hence in every quadrilateral containing three right angles, and one obtuse, or acute, the sides adjacent to this oblique angle are less respectively than the opposite sides if this angle is obtuse, but greater if it is acute.

For this has just now been demonstrated of the side CH relatively to the opposite side AM; in the same way it is demonstrated of the side AC relatively to the opposite side MH. For since the straights AC, MH, are perpendicular to this AM, they cannot (P. I.) be mutually equal, on account of the unequal angles at the join CH.

But neither (in the hypothesis of an obtuse angle at

dam AN, portio ipsius AC, aequalis esse ipsi MH, qua nimirum major sit praedicta AC: caeterum (ex eadem prima) aequales forent anguli ad junctam HN; quod est absurdum, ut supra. Rursum vero (in hypothesi anguli acuti in eo puncto C) si velis quandam AX, sumptam in AC protracta, aequalem ipsi MH, qua nimirum minor sit modo dicta AC; jam eodem titulo aequales erunt anguli ad HX; quod utique absurdum itidem est, ut supra. Restat igitur, ut in hypothesi quidem anguli obtusi in eo puncto C, latus AC minus sit contraposito latere MH; in hypothesi autem anguli acuti sit eodem majus. Quod erat intentum.

COROLLARIUM II.

Multo autem magis erit CH major portione qualibet ipsius AM, ut puta PM, ad quam nempe juncta [5] CP acutiorem adhuc angulum efficiat cum ipso CH versus partes puncti H, et obtusum (ex decimasexta primi) cum ea PM versus partes puncti M.

COROLLARIUM III.

Rursum constat praedicta omnia aeque procedere, sive assumpta perpendicula AC, et BD, fuerint certae cujusdam apud nos longitudinis, sive sint, aut supponantur infinite parva. Quod quidem notari opportune debet in reliquis sequentibus Propositionibus.

PROPOSITIO IV.

Vicissim autem (manente figura praecedentis Propositionis) anguli ad junctam CD erunt aut recti, aut obtusi, aut acuti, prout recta CD aequalis fuerit, aut minor, aut major, contraposita AB.

C) can a certain AN, a piece of this AC, than which certainly the aforesaid AC is greater, be equal to this MH: otherwise (P. I.) the angles at the join HN would be equal; which is absurd, as above.

Again however (in the hypothesis of an acute angle at this point C), if you take a certain AX, assumed on AC produced, than which certainly the just mentioned AC is less, equal to this MH; now by this same title the angles at HX will be equal; which assuredly is absurd in the same way, as above.

It remains therefore, that indeed in the hypothesis of an obtuse angle at this point C, the side AC is less than the opposite side MH; but in the hypothesis of an acute angle is greater than it.

Quod erat intentum.

COROLLARY II. ✲

But by much more will CH be greater than any piece of this AM, as for instance PM, since of course the join [5]CP makes an angle still more acute with this CH toward the parts of the point H, and obtuse (Eu. I. 16) with this PM toward the parts of the point M.

COROLLARY III.

Again it abides that all things aforesaid equally result, whether the assumed perpendiculars AC, and BD are of some length fixed by us, are, or are supposed infinitesimal.

This indeed ought opportunely to be noted in remaining subsequent propositions.

PROPOSITION IV.

But inversely (the figure of the preceding proposition remaining) the angles at the join CD will be right, or obtuse, or acute, according as the straight CD is equal, or less, or greater than the opposite AB.

Demonstratur. Si enim recta CD aequalis sit contrapositae AB, et nihilominus anguli ad eandem sint aut obtusi, aut acuti; jam ipsi tales anguli eam probabunt (ex praecedente) non aequalem, sed minorem, aut majorem contraposita AB; quod est absurdum contra hypothesim. Idem uniformiter valet circa reliquos casus. Stat igitur angulos ad junctam CD esse aut rectos, aut obtusos, aut acutos, prout recta CD aequalis fuerit, aut minor, aut major contraposita AB. Quod erat demonstrandum.

DEFINITIONES.

Quandoquidem (ex prima hujus) recta jungens extremitates aequalium perpendiculorum eidem rectae (quam vocabimus basim) insistentium, aequales ef-[6]ficit angulos cum ipsis perpendiculis; tres idcirco distinguendae sunt hypotheses circa speciem horum angulorum. Et primam quidem appellabo hypothesim anguli recti; secundam vero, et tertiam appellabo hypothesim anguli obtusi, et hypothesim anguli acuti.

PROPOSITIO V.

Hypothesis anguli recti, si vel in uno casu est vera, semper in omni casu illa sola est vera.

Demonstratur. Efficiat juncta CD (fig. 4.) angulos rectos cum duobus quibusvis aequalibus perpendiculis AC. BD, uni cuivis AB insistentibus. Erit CD (ex tertia hujus) aequalis ipsi AB. Sumantur in AC, et BD protractis duae CR, DX, aequales ipsis AC, BD; jungaturque RX. Facile ostendemus junctam RX aequalem fore ipsi AB, et angulos ad eandem rectos. Et primo quidem per

PROOF. For if the straight CD is equal to the opposite AB, and nevertheless the angles at it are either obtuse, or acute; now these such angles prove it (P. III.) not equal, but less, or greater than the opposite AB; which is absurd against the hypothesis.

The same uniformly avails in regard to the remaining cases. It holds therefore that the angles at the join CD are either right, or obtuse, or acute, according as the straight CD is equal to, or less, or greater than the opposite AB.

Quod erat demonstrandum.

DEFINITIONS.

Since (P. I.) the straight joining the extremities of equal perpendiculars standing upon the same straight (which we call base), makes equal [6] angles with these perpendiculars; therefore there are three hypotheses to be distinguished according to the species of these angles. And the first indeed I will call hypothesis of right angle; the second however, and the third I will call hypothesis of obtuse angle, and hypothesis of acute angle.

PROPOSITION V.

If even in a single case the hypothesis of right angle is true, always in every case it alone is true.

PROOF. Let the join CD (fig. 4) make right angles with any two perpendiculars AC, BD, standing upon any straight AB.

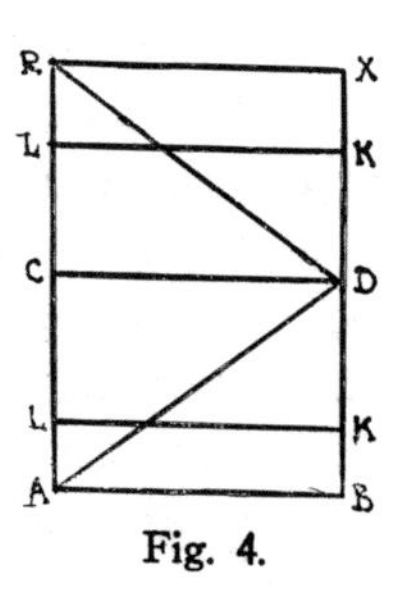

Fig. 4.

CD will be equal to this AB. Assume in AC, and BD produced two sects CR, DX, equal to these AC, BD; and join RX. We may easily show that the join RX will be equal to this AB, and the angles at it right. And first indeed by

superpositionem quadrilateri ABDC super quadrilaterum CDXR, adhibita communi basi CD. Deinde elegantius sic proceditur. Jungantur AD, RD. Constat (ex quarta primi) aequales fore in triangulis ACD, RCD, bases AD, RD, atque item angulos CDA, CDR, ac propterea aequales reliquos ad unum rectum, nimirum ADB, RDX. Quare rursum (ex eadem quarta primi) aequalis erit, in triangulis ADB, RDX, basis AB, basi RX. Igitur (ex praecedente) anguli ad junctam RX erunt recti, ac propterea persistemus in eadem hypothesi anguli recti.

Quoniam vero augeri similiter potest longitudo perpendiculorum in infinitum, sub eadem basi AB, consistente semper hypothesi anguli recti, demonstrandum est eandem hypothesim semper mansuram in casu cujusvis imminutionis eorundem perpendiculorum; quod quidem ita evincitur. [7]

Sumantur in AR, et BX duo quaelibet aequalia perpendicula AL, BK, jungaturque LK. Si anguli ad junctam LK recti non sint, erunt tamen (ex prima hujus) invicem aequales. Erunt igitur ex una parte, ut puta versus AB obtusi, et versus RX acuti, ut nimirum anguli hinc inde ad utrunque illorum punctorum aequales sint (ex decimatertia primi) duobus rectis. Constat autem aequalia etiam invicem esse perpendicula LR, KX, ipsi RX insistentia. Igitur (ex tertia hujus) erit LK major quidem contraposita RX, et minor contraposita AB.

Hoc autem absurdum est; cum AB, et RX ostensae sint aequales. Non ergo mutabitur hypothesis anguli recti sub quacunque imminutione perpendiculorum, dum consistat semel posita basis AB.

Sed neque immutabitur hypothesis anguli recti, sub quacunque imminutione, aut majori amplitudine basis; cum manifestum sit considerari posse ut basim quodvis

superposition of the quadrilateral ABDC upon the quadrilateral CDXR, applied to the common base CD.

Also we may proceed more elegantly thus. Join AD, RD. It follows (Eu. I. 4) in the triangles ACD, RCD, the bases AD, RD will be equal and likewise the angles CDA, CDR, and certainly ADB, RDX because equal remainders from a right angle. Whereby in turn (Eu. I. 4) in the triangles ADB, RDX, the base AB will be equal to the base RX. Therefore (P. IV.) the angles at the join RX will be right, and so we abide in the same hypothesis of right angle. ✡

Since now the length of the perpendiculars can be similarly increased infinitely, under the same base AB, ✡ the hypothesis of right angle always subsisting, it only remains to be proved that the same hypothesis will always abide in any case of diminution of those perpendiculars; which indeed is thus evinced. [7]

Assume in AR, and BX any two equal perpendiculars AL, BK, and join LK. If the angles at the join LK are not right, nevertheless (P. I.) they will be equal to each other. Therefore they will be toward one part, as suppose toward AB obtuse, and toward RX acute, since certainly the angles here at each of those points are (Eu. I. 13) equal to two rights.

But it also holds that the perpendiculars LR, KX, those standing upon RX, will be mutually equal. Therefore (P. III.) LK will be greater indeed than the opposite RX, and less than the opposite AB.

But this is absurd; because AB, and RX have been shown equal. Therefore the hypothesis of right angle is not changed by any diminution of the perpendiculars, whilst abides the once posited base AB.

But neither is the hypothesis of right angle changed for any diminution, or greater amplitude of the base; since manifestly may be considered as base any perpen-

perpendiculum BK, aut BX, atque ideo considerari vicissim ut perpendicula ipsam AB, et rectam aequalem contrapositam KL, aut XR.

Constat igitur hypothesim anguli recti, si vel in uno casu sit vera, semper in omni casu illam solam esse veram. Quod erat demonstrandum.

PROPOSITIO VI.

Hypothesis anguli obtusi, si vel in uno casu est vera, semper in omni casu illa sola est vera.

Demonstratur. Efficiat juncta CD (fig. 5.) angulos obtusos cum duobus quibusvis aequalibus perpendiculis AC, BD, uni cuivis rectae AB insistentibus. Erit CD (ex tertia hujus) minor ipsa AB. Sumantur in AC, BD protractis duae quaelibet invicem aequales portiones CR, [8] DX; jungaturque RX. Jam quaero de angulis ad junctam RX, qui utique (ex prima hujus) aequales invicem erunt. Si obtusi sunt, habemus intentum. At recti non sunt; quia sic unum haberemus casum pro hypothesi anguli recti, qui nullum (ex praecedente) relinqueret locum pro hypothesi anguli obtusi. Sed neque acuti sunt. Nam sic esset RX (ex tertia hujus) major ipsa AB; ac propterea multo major ipsa CD. Hoc autem subsistere non posse sic ostenditur. Si quadrilaterum CDXR intelligatur impleri rectis abscindentibus ab ipsis CR, DX, portiones invicem aequales, implicat transiri a recta CD, quae minor est ipsa AB, ad RX eadem majorem, quin

dicular BK, or BX, and therefore may be considered in turn as perpendiculars that AB, and the equal opposite sect KL, or XR.

Therefore is established that if even in a single case the hypothesis of right angle be true, always in every case it alone is true.

Quod erat demonstrandum.

PROPOSITION VI.

If even in a single case the hypothesis of obtuse angle is true, always in every case it alone is true.

PROOF. Let the join CD (fig. 5) make obtuse angles with any two equal perpendiculars AC, BD, standing upon any straight AB.

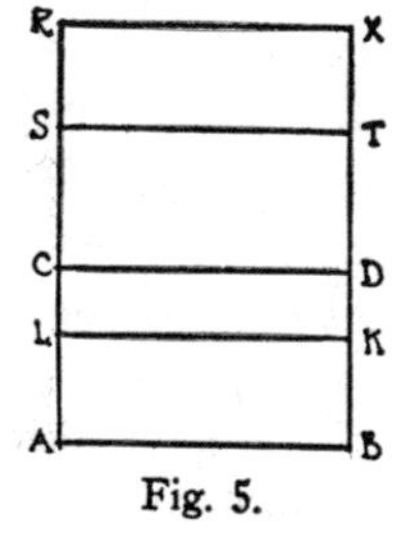

Fig. 5.

CD will be (P. III.) less than this AB.

Assume in AC and BD produced any two mutually equal portions CR [8] and DX; and join RX.

Now I investigate the angles at the join RX, which certainly (P. I.) will be mutually equal.

If they are obtuse we have our assertion.

But they are not right; because thus we would have a case for the hypothesis of right angle, which (P. V.) would leave no place for the hypothesis of obtuse angle. But neither are they acute.

For thus RX would be (P. III.) greater than this AB; and still more therefore greater than CD itself. But that this cannot be is thus shown. If the quadrilateral CDXR is taken to be filled up by straights cutting off from these CR, DX, portions mutually equal, this implies transition from the sect CD, which is less than AB itself, to RX greater than it, verily transition through a certain

transeatur per quandam ST ipsi AB aequalem. Hoc autem absurdum esse in hac hypothesi ex eo constat; quia sic (ex quarta hujus) unus haberetur casus pro hypothesi anguli recti, qui nullum (ex praecedente) relinqueret locum hypothesi anguli obtusi. Igitur anguli ad junctam RX debent esse obtusi.

Deinde, sumptis in AC, BD, aequalibus portionibus AL, BK; simili modo ostendemus angulos ad junctam LK nequire esse acutos versus ipsam AB; quia sic illa foret major, quam AB, ac propterea multo major recta CD. Hinc autem reperiri deberet, ut supra, quaedam intermedia inter CD minorem, et LK majorem ipsa AB intermedia, inquam, aequalis ipsi AB, quae utique, ex jam notis, omnem locum auferret hypothesi anguli obtusi. Tandem propter hanc ipsam causam recti esse nequeunt anguli ad junctam LK; ergo erunt obtusi. Igitur sub eadem basi AB, auctis, aut imminutis ad libitum perpendiculis, manebit semper hypothesis anguli obtusi.

Sed debet idem demonstrari sub assumpta qualibet basi. Eligatur (fig. 6.) pro basi quodlibet ex praedictis perpendiculis, ut puta BX. Dividantur bifariam in punctis [9] M, et H ipsae AB, RX; jungaturque MH. Erit MH (ex secunda hujus) perpendicularis ipsis AB, RX. Est autem angulus ad punctum B rectus ex hypothesi; et obtusus, ex jam demonstratis, ad punctum X. Fiat igitur angulus rectus BXP versus partes ipsius MH. Occurret XP ipsi MH in quodam puncto P inter puncta M, et H

ST equal to this AB.✲ But that this is absurd in the present hypothesis follows so; because thus (P. IV.) we have a case for the hypothesis of right angle, which (P. V.) would leave no place for the hypothesis of obtuse angle. Therefore the angles at the join RX must be obtuse.

Then, equal portions AL, BK being assumed in AC, BD; in a similar manner we show the angles at the join LK cannot be acute toward this AB; because thus it would be greater than AB, and still more therefore greater than the sect CD. But here would be found, as above, a certain intermediate between CD less, and LK greater than this AB; an intermediate, I say, equal to AB itself, which certainly, from what was just now observed, would take away every place for the hypothesis of obtuse angle.

Finally from this very cause the angles at the join LK cannot be right; therefore they will be obtuse.

Theretore with the same base AB, the perpendiculars being increased or diminished at will, the hypothesis of obtuse angle will always persist.

But the same ought to be demonstrated for any assumed base.

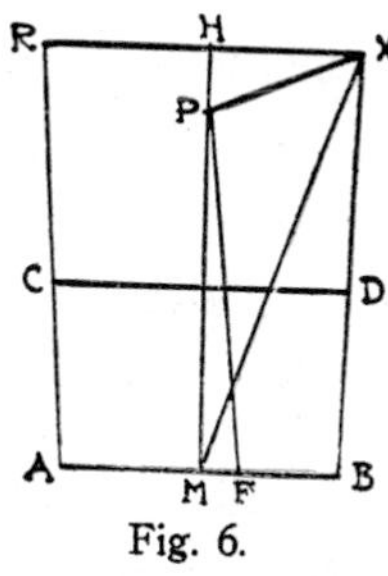

Fig. 6.

Let there be chosen (fig. 6) for base any one of the aforesaid perpendiculars, as BX suppose.

Let AB, RX be bisected in the points [9] M and H; and MH joined. MH will be (P. II.) perpendicular to AB, RX. But the angle at the point B is right by hypothesis; and at the point X obtuse, from what has just now been demonstrated.

Make therefore the right angle BXP toward the parts of this MH. XP will meet MH itself in some point P situated between the points M and H; since on the one

constituto; cum ex una parte angulus BXH sit obtusus; et ex altera, si jungatur XM, angulus BXM (ex decimaseptima primi) sit acutus. Tum vero; quoniam quadrilaterum XBMP tres continet angulos rectos ex jam notis, et unum obtusum (ex decimasexta primi) in puncto P, quia est externus relate ad internum, et oppositum rectum angulum in puncto H trianguli PHX; erit latus XP (ex Cor. I. post tertiam hujus) minus contraposito BM. Quare; assumpta in BM portione BF aequali ipsi XP; erunt (ex prima hujus) anguli ad junctam PF invicem aequales, nimirum obtusi, cum angulus BFP (ex decimasexta primi) sit obtusus propter rectum angulum internum, et oppositum FMP. Igitur sub qualibet basi BX consistit hypothesis anguli obtusi.

Consistet autem, ut supra, eadem hypthesis sub eadem basi BX, quamvis aequalia perpendicula ad libitum augeantur, aut minuantur. Itaque constat hypothesim anguli obtusi, si vel in uno casu sit vera, semper in omni casu illam solam esse veram. Quod erat demonstrandum.

PROPOSITIO VII.

Hypothesis anguli acuti, si vel in uno casu est vera, semper in omni casu illa sola est vera.

Demonstratur facillime. Si enim hypothesis anguli acuti permittat aliquem casum alterutrius hypothesis aut anguli recti, aut anguli obtusi, jam (ex duabus praecеden-[10]tibus) nullus relinquetur locus ipsi hypothesi anguli acuti; quod est absurdum. Itaque hypothesis anguli acuti, si vel in uno casu est vera, semper in omni casu illa sola est vera. Quod erat demonstrandum.

hand the angle BXH is obtuse; and, on the other, if XM be joined, the angle BXM (Eu. I. 17) is acute. Then however, since the quadrilateral XBMP contains three right angles, from what has just now been noted, and one obtuse (Eu. I. 16) at the point P, because it is external in relation to the internal and opposite right angle at the point H of the triangle PHX; the side XP will be (Cor. I., P. III.) less than the opposite BM. Wherefore, assuming in BM the portion BF equal to this XP, the angles at the join PF will be (P. I.) mutually equal, certainly obtuse, since the angle BFP (Eu. I. 16) is obtuse because of the right angle interior and opposite FMP. Therefore the hypothesis of obtuse angle abides for any base BX.

But, as above, this hypothesis abides for this base BX, however much the equal perpendiculars are augmented or diminished at will. Therefore it holds, that if even in a single case the hypothesis of obtuse angle is true, always in every case it alone is true.

Quod erat demonstrandum.

PROPOSITION VII.

If even in a single case the hypothesis of acute angle is true, always in every case it alone is true.

Proof is very easily given. For if the hypothesis of acute angle should permit any case of either other hypothesis, either of right angle, or of obtuse angle, now (from the two preceding [10] propositions) no place would be left for the hypothesis of acute angle; which is absurd.

Therefore if even in a single case the hypothesis of acute angle is true, always in every case it alone is true.

Quod erat demonstrandum.

PROPOSITIO VIII.

Dato quovis triangulo (fig. 7.) ABD, rectangulo in B, protrahatur DA usque ad aliquod punctum X, et per A erigatur ipsi AB perpendicularis HAC, existente puncto H ad partes anguli XAB. Dico angulum externum XAH aequalem fore, aut minorem, aut majorem interno, et opposito ADB, prout vera sit hypothesis anguli recti, aut anguli obtusi, aut anguli acuti: Et vicissim.

Demonstratur. Sumatur in HC portio AC aequalis ipsi BD, jungaturque CD. Erit CD, in hypothesi anguli recti, aequalis (ex tertia hujus) ipsi AB. Quare angulus ADB aequalis erit (ex octava primi) angulo DAC, sive ejus aequali (ex decimaquinta primi) angulo XAH. Quod erat primo loco demonstrandum.

Tum, in hypothesi anguli obtusi, erit CD (ex eadem tertia hujus) minor ipsa AB. Quare in triangulis ADB, DAC erit (ex vigesimaquinta primi) angulus DAC, sive (ipsi ad verticem) XAH, minor angulo ADB. Quod erat secundo loco demonstrandum.

Tandem, in hypothesi anguli acuti, erit CD (ex eadem tertia hujus) major contraposita AB. Quare in praedictis triangulis, erit (ex eadem vigesimaquinta primi) angulus DAC, sive (ipsi ad verticem) XAH, major angulo ADB. Quod erat tertio loco demonstrandum.

Vicissim autem: si angulus CAD, sive ejus ad verticem XAH, aequalis sit interno, et opposito ADB; erit (ex quarta primi) juncta CD aequalis ipsi AB, ac propte-[11]rea (ex quarta hujus) vera erit hypothesis anguli recti.

PROPOSITION VIII.

Given any triangle (fig. 7) ABD, right-angled at B; prolong DA to any point X, and through A erect HAC perpendicular to AB, the point H being within the angle XAB. I say the external angle XAH will be equal to, or less, or greater than the internal and opposite ADB, according as is true the hypothesis of right angle, or obtuse angle, or acute angle: and inversely.

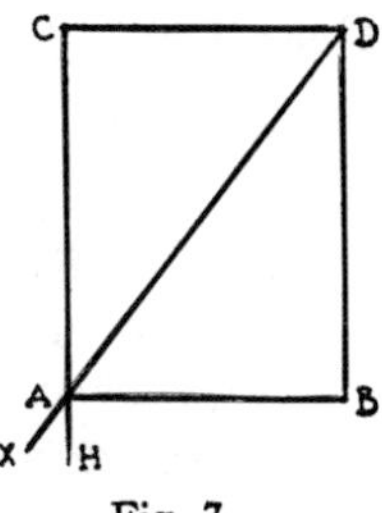

Fig. 7.

Proof. Assume in HC the portion AC equal to BD, and join CD. CD will be, in the hypothesis of right angle (P. III.) equal to AB. Wherefore the angle ADB will be equal (Eu. I. 8) to the angle DAC, or to its equal (Eu. I. 15) the angle XAH.

Quod erat primo loco demonstrandum.

Then, in the hypothesis of obtuse angle, CD will be (P. III.) less than AB.

Wherefore in the triangles ADB, DAC the angle DAC, or its vertical XAH, will be (Eu. I. 25) less than the angle ADB.

Quod erat secundo loco demonstrandum.

Finally, in the hypothesis of acute angle, CD will be (P. III.) greater than the opposite AB. Wherefore in the said triangle the angle DAC, or its vertical XAH, will be (Eu. I. 25) greater than the angle ADB.

Quod erat tertio loco demonstrandum.

But inversely: if the angle CAD, or its vertical XAH, be equal to the internal and opposite ADB; the join CD will be (Eu. I. 4) equal to AB, and therefore [11] the hypothesis of right angle will be (P. IV.) true.

Sin vero angulus CAD, sive ejus ad verticem XAH, minor sit, aut major interno, et opposito ADB; erit etiam (ex vigesimaquarta primi) juncta CD minor, aut major ipsa AB; ac propterea (ex quarta hujus) vera erit respective hypothesis aut anguli obtusi, aut anguli acuti Quae omnia erant demonstranda.

PROPOSITIO IX.

Cujusvis trianguli rectanguli reliqui duo acuti anguli simul sumpti aequales sunt uni recto, in hypothesi anguli recti; majores uno recto, in hypothesi anguli obtusi; minores autem in hypothesi anguli acuti.

Demonstratur. Si enim angulus XAH (manente figura superioris Propositionis) aequalis est (nimirum, ex praecedente, in hypothesi anguli recti) angulo ADB; jam angulus ADB duos rectos efficiet cum angulo HAD, prout eos efficit (ex decimatertia primi) praedictus angulus XAH cum eodem angulo HAD. Quare, dempto recto angulo HAB, aequales manebunt uni recto duo simul anguli ADB, et BAD. Quod erat primum.

Tum vero; si angulus XAH minor est (nimirum, ex praecedente, in hypothesi anguli obtusi) angulo ADB, jam angulus ADB plusquam duos rectos efficiet cum angulo HAD, cum quo duos efficit rectos (ex praedicta decimatertia primi) angulus XAH. Quare, dempto angulo HAB, majores erunt uno recto duo simul anguli ADB, et BAD. Quod erat secundum.

Tandem, si angulus XAH major sit (nimirum, ex praecedente, in hypothesi anguli acuti) angulo ADB; jam angulus ADB minus quam duos rectos efficiet cum angulo HAD, cum quo duos efficit rectos (ex eadem decima-

But if however the angle CAD, or its vertical XAH, be less, or greater than the internal or opposite ADB; also the join CD will be (Eu. I. 24) less or greater than AB; and therefore (P. IV.) will be true respectively the hypothesis of obtuse angle, or acute angle.

Quae omnia erant demonstranda.

PROPOSITION IX.

In any right-angled triangle the two acute angles remaining are, taken together, equal to one right angle, in the hypothesis of right angle; greater than one right angle, in the hypothesis of obtuse angle; but less in the hypothesis of acute angle. ✡

PROOF. For if the angle XAH (fig. 7) is equal to the angle ADB, which is certain from the preceding proposition in the hypothesis of right angle, then the angle ADB makes up with the angle HAD two right angles, as (Eu. I. 13) the aforesaid angle XAH makes them up with this angle HAD. Wherefore, the right angle HAB being subtracted, the two angles ADB and BAD remain together equal to one right angle.

Quod erat primum.

However, if the angle XAH is less than the angle ADB, which is certain from the preceding proposition in the hypothesis of obtuse angle, then the angle ADB makes up with the angle HAD more than two right angles, since with it (Eu. I. 13) the angle XAH makes up two. Wherefore, the angle HAB being subtracted, the two angles ADB and BAD will be together greater than one right angle.

Quod erat secundum.

Finally, if the angle XAH be greater than the angle ADB, which is certain from the preceding proposition in the hypothesis of acute angle, then the angle ADB will make up less than two right angles with the angle HAD,

[12]tertia primi) angulus XAH. Quare, dempto angulo recto HAB, minores erunt uno recto duo simul anguli ADB, et BAD. Quod erat tertium.

PROPOSITIO X.

Si recta DB (fig. 8.) perpendiculariter insistat cuidam ABM, sitque juncta DM major juncta DA, etiam basis BM major erit basi BA. Et vicissim.

Demonstratur. Et primo quidem non erunt illae bases invicem aequales. Caeterum (ex quarta primi) aequales forent, contra hypothesim, ipsae AD, DM. Sed neque erit BA major quam BM. Caeterum, sumpta in BA portione BS aequali ipsi BM, junctaque SD, aequales forent (ex eadem quarta primi) anguli BSD, BMD: Est autem angulus BSD (ex decimasexta primi) major angulo BAD. Ergo eodem major foret angulus BMD. Hoc autem est contra decimamoctavam primi; cum latus DM in triangulo MDA supponatur majus latere DA. Restat igitur, ut basis BM major sit basi BA. Quod erat primo loco demonstrandum.

Deinde si alterutra basis, ut puta BA (ne immutetur figura) fingatur major altera BM; tunc juncta DS, quae ex BA abscindat portionem SB aequalem ipsi BM, aequalis erit (ex quarta primi) junctae DM. Rursum obtusus erit (ex decimasexta primi) angulus DSA, et acutus (ex decimaseptima ejusdem primi) angulus DAS. Quare (ex

since with this (Eu. I. 13) [12] the angle XAH makes up two. Wherefore, subtracting the right angle HAB, the angles ADB and BAD will be together less than one right angle.

Quod erat tertium.

PROPOSITION X.

If the straight DB (fig. 8) stand perpendicular to a straight ABM, and the join DM be greater than the join DA, then also the base BM will be greater than the base BA. And inversely.

PROOF. And in the first place assuredly these bases will not be mutually equal. Otherwise (Eu. I. 4) AD and DM would be equal, contrary to the hypothesis.

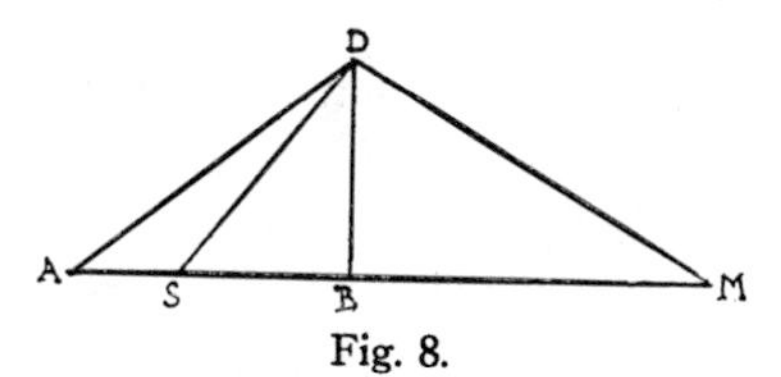

Fig. 8.

But neither will BA be greater than BM. Otherwise, in BA the portion BS being taken equal to BM, and SD joined, the angles BSD, BMD (Eu. I. 4) would be equal. But angle BSD is (Eu. I. 16) greater than angle BAD. Therefore angle BMD would be greater than angle BAD. But this is contrary to Eu. I. 18; since side DM in triangle MDA is supposed greater than side DA. It remains therefore, that the base BM is greater than the base BA.

Quod erat primo loco demonstrandum.

Next if either base, as BA suppose (the figure need not be changed) is conceived as greater than the other BM; then the join DS, which cuts off from BA the portion SB equal to BM, will be equal (Eu. I. 4) to the join DM. Again angle DSA will be obtuse (Eu. I. 16) and angle DAS acute (Eu. I. 17).

decimanona ejusdem) erit juncta DA major juncta DS, ejusque supposita aequali juncta DM. Quod erat secundo loco demonstrandum. Itaque constant proposita.[13]

PROPOSITIO XI.

Recta AP (quantaelibet longitudinis) secet duas rectas PL, AD (fig. 9.) priorem quidem sub recto angulo in P, posteriorem vero in A sub quovis acuto angulo convergente ad partes ipsius PL. Dico rectas AD, PL (in hypothesi anguli recti) in aliquo puncto, et quidem ad finitam, seu terminatam distantiam, tandem coituras, si protrahantur versus illas partes, ad quas cum subjecta AP duos angulos efficiunt duobus rectis minores.

Demonstratur. Protrahatur DA versus alias partes usque ad aliquod punctum X, et per A erigatur ipsi AP perpendicularis HAC, existente puncto H ad partes anguli XAP. Tum in AD protracta versus partes ipsius PL sumantur duo aequalia intervalla AD, DF, demittanturque ad subjectam AP perpendiculares DB, FM, quae utique cadent, propter decimamseptimam primi, ad partes

Wherefore (Eu. I. 19) the join DA will be greater than the join DS, and the join supposed equal to it DM.

Quod erat secundo loco demonstrandum. Itaque constant proposita. [13]

PROPOSITION XI.

Let the straight AP (as long as you choose) cut the two straights PL, AD (fig. 9), the first indeed at right angles in P, but the latter at A in any acute angle converging toward the parts PL. I say the straights AD, PL (in the hypothesis of right angle) will at length meet in some point, and indeed at a finite, or terminated distance, if they are prolonged toward those parts on which they make with the transversal AP two angles together less than two right angles.

PROOF. Prolong DA toward the other parts even to some point X, and through A erect to AP the perpendicular HAC, the point H being toward the parts of the angle XAP.

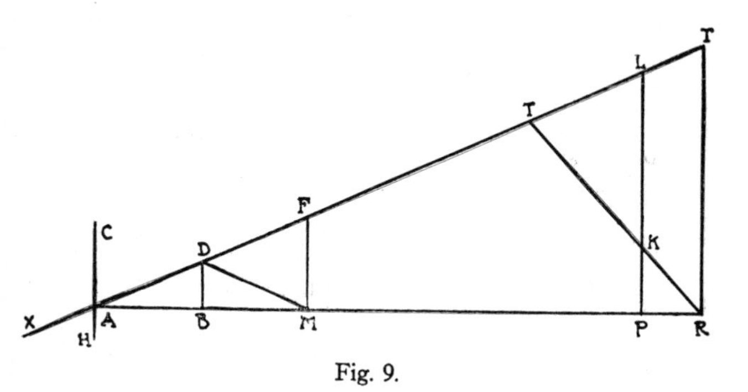

Fig. 9.

Then in AD produced toward the parts of PL assume two equal intervals AD, DF, and let fall upon the transversal AP the perpendiculars DB, FM, which certainly (Eu. I. 17) fall toward the parts of the acute angle DAP;

anguli acuti DAP; jungaturque DM. Ostendere debeo junctam DM aequalem fore ipsi DF, sive DA.

Et primo quidem nequit DM major esse ipsa DF. Caeterum enim angulus DMF minor foret (ex decima-octava primi) angulo DFM, sive ejus aequali (ex octava hujus, in hypothesi anguli recti) angulo XAH, sive ejus ad verticem CAD. Quare (cum anguli CAM, FMA ponantur aequales, utpote recti) reliquus angulus DMA major foret reliquo angulo DAM. Hoc autem absurdum est (contra decimamoctavam primi) si nempe DM ponatur major ipsa DF, sive DA.

Sed neque erit DM minor ipsa DF. Caeterum angulus DMF major foret (ex eadem decimaoctava primi) angulo DFM, sive ejus aequali (ex praedicta octava hujus, in hypothesi anguli recti) angulo XAH, sive ejus ad verticem CAD. Quare rursum, ut supra, reliquus angulus [14] DMA non major, sed minor foret reliquo angulo DAM. Hoc autem absurdum est (contra eandem decimamoctavam primi) si nempe DM ponatur minor ipsa DF, sive DA.

Restat igitur, ut juncta DM aequalis sit ipsi DF, sive DA. Quare in triangulo DAM aequales erunt (ex quinta primi) anguli ad puncta A, et M; atque ideo in triangulis DBA, DBM, rectangulis in B, aequales erunt (ex vigesimasexta primi) bases AB, BM. Quod quidem hoc loco intendebatur.

Quoniam igitur (assumpto in AD continuata intervallo AF duplo intervalli AD) perpendicularis FM ad subjectam AP demissa abscindit ex AP versus P basim AM duplam illius AB, quam abscindit perpendicularis demissa ex puncto D; manifestum est tot vicibus fieri posse hanc praecedentis intervalli duplicationem, ut sic in ipsa AD continuata deveniatur ad quoddam punctum T, ex quo perpendicularis demissa ad continuatam AP abscindat quandam AR majorem ipsa quantalibet finita AP.

and join DM. I should show that the join DM will be equal to DF, or DA.

And in the first place indeed DM cannot be greater than DF. For otherwise the angle DMF would be less (Eu. I. 18) than the angle DFM, or its equal (P. VIII., in the hypothesis of right angle) the angle XAH, or its vertical CAD. Wherefore (since the angles CAM, FMA are assumed equal, as being right) the remaining angle DMA would be greater than the remaining angle DAM. But this is absurd (against Eu. I. 18) if indeed DM is taken greater than DF or DA.

But neither will DM be less than this DF. Otherwise the angle DMF would be greater (Eu. I. 18) than the angle DFM, or its equal (P. VIII., in hypothesis of right angle) the angle XAH, or its vertical CAD. Wherefore again, as above, the remaining angle [14] DMA will not be greater, but less than the remaining angle DAM. But this is absurd (against Eu. I. 18) if indeed DM is taken less than DF, or DA.

It remains therefore, that the join DM is equal to DF, or DA. Wherefore in the triangle DAM (Eu. I. 5) the angles at the points A, and M will be equal; and therefore in the triangles DBA, DBM, right-angled at B, the bases AB, BM will be equal (Eu. I. 26). This indeed was here our aim.

Since therefore (assuming in AD produced the interval AF double the interval AD) the perpendicular FM let fall on the transversal AP cuts off from AP toward P a base AM double AB, which the perpendicular let fall from the point D cuts off; it is manifest that this duplication of the preceding interval can be so many times repeated, that thus in AD continued we attain to a certain point T, from which the perpendicular let fall upon AP prolonged cuts off a certain AR greater than the finite AP however great.

Constat autem evenire id non posse, nisi post occursum ipsius continuatae AD in quoddam punctum L ipsius PL. Si enim punctum T consisteret ante illum occursum, deberet ipsa perpendicularis TR secare eandem PL in quodam puncto K. Tunc autem in triangulo KPR invenirentur duo anguli recti in punctis P, et R; quod est absurdum contra decimamseptimam primi. Itaque constat rectas AD, PL sibi invicem (in hypothesi anguli recti) in aliquo puncto occursuras (et quidem ad finitam, seu terminatam distantiam) si protrahantur versus illas partes, ad quas cum subjecta AP (quantaelibet finitae longitudinis) duos angulos efficiunt duobus rectis minores. Quod erat demonstrandum. [15]

PROPOSITIO XII.

Rursum dico rectam AD alicubi ad eas partes occursuram rectae PL (et quidem ad finitam, seu terminatam distantiam) etiam in hypothesi anguli obtusi.

Demonstratur. Nam sumpta, ut in superiore Propositione, DF aequali ipsi AD, demissisque jam notis perpendicularibus, ostendere debeo junctam DM majorem fore ipsa DF, sive DA, atque ideo (ex decima hujus) rectam BM majorem fore ipsa AB. Et primo non erit DM aequalis ipsi DF. Caeterum angulus DMF aequalis foret

But it is evident this cannot happen, except after the meeting of the prolonged AD with PL in some point L.

For if the point T occurred before that meeting, the perpendicular TR must cut PL in some point K. But then in the triangle KPR would be found two right angles at the points P and R; which is absurd (against Eu. I. 17).

Therefore it holds that the straights AD, PL meet each other mutually (in the hypothesis of right angle) in some point (and indeed at a finite or terminated distance) if they be produced toward that side, on which with the transversal AP (of finite length as great as you choose) they make two angles together less than two right angles.

Quod erat demonstrandum. [15]

PROPOSITION XII. ✡

Again I say also in the hypothesis of obtuse angle the straight AD will meet the straight PL somewhere toward those parts (and indeed at a finite, or terminated distance).

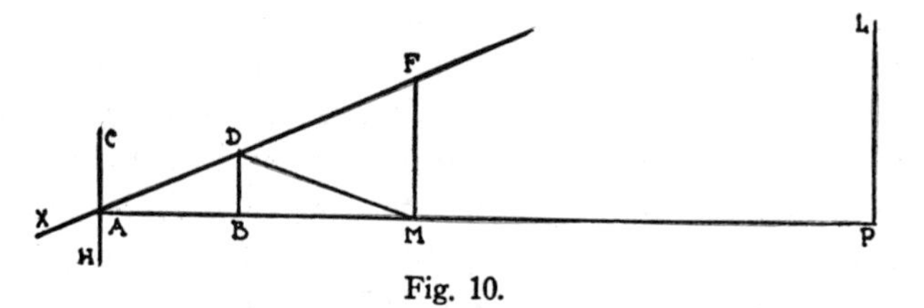

Fig. 10.

PROOF. For, as in the preceding proposition, DF being assumed equal to AD, and the just noted perpendiculars let fall, I must show the join DM will be greater than DF, or DA, and therefore (P. X.) the straight BM will be greater than AB.

And in the first place DM will not be equal to DF. Otherwise the angle DMF would be equal (Eu. I. 5) to

(ex quinta primi) angulo DFM, atque ideo major (ex octava hujus in hypothesi anguli obtusi) angulo externo XAH, sive ejus ad verticem CAF. Quare (cum anguli CAM, FMA ponantur aequales utpote recti) reliquus angulus DMA minor foret reliquo angulo DAM. Quod est absurdum contra quintam primi, si nempe DM aequalis sit ipsi DF, sive DA.

Sed neque ipsa DM minor est altera DF, sive DA. Caeterum (ex decimaoctava primi) angulus DMF major foret angulo DFM, atque ideo (in hac hypothesi anguli obtusi) multo major angulo externo XAH, sive ejus ad verticem CAD. Quare rursum, ut supra, reliquus angulus DMA multo minor foret reliquo angulo DAM. Hoc autem absurdum est, contra eandem decimamoctavam primi, si nempe DM minor sit ipsa DF, sive DA.

Restat igitur, ut juncta DM major sit ipsa DF, sive DA, atque ideo (ex decima hujus) ipsa BM major sit altera AB. Quod erat hoc loco intentum.

Quoniam igitur, assumpto in AD continuata intervallo AF duplo intervalli AD, perpendicularis FM ad subjectam AP demissa plus duplo ex eadem abscindit, quam abscindatur a perpendiculari demissa ex puncto D; [16] multo citius in hac hypothesi anguli obtusi, quam in superiore hypothesi anguli recti, devenietur ad tantum intervallum, ex quo perpendicularis demissa abscindat basim majorem ipsa quantalibet designata AP. Hoc autem, ut in superiore Propositione, contingere nequit, nisi post occursum continuatae AD in aliquod punctum ipsius PL; et quidem ad finitam, seu terminatam distantiam. Quod erat etc.

the angle DFM, and therefore greater (P. VIII., in the hypothesis of obtuse angle) than the external angle XAH, or its vertical CAF.

Wherefore (since the angles CAM, FMA are taken equal, as being right) the remaining angle DMA would be less than the remaining angle DAM. This is absurd (against Eu. I. 5), if indeed DM be equal to DF, or DA.

But neither is DM less than DF, or DA. Otherwise (Eu. I. 18) the angle DMF would be greater than the angle DFM, and therefore still greater (in the hypothesis of obtuse angle) than the external angle XAH, or its vertical CAD. Wherefore again, as above, the remaining angle DMA would be still less than the remaining angle DAM. But this is absurd (against Eu. I. 18) if indeed DM be less than DF, or DA.

It remains therefore, that the join DM is greater than DF, or DA, and therefore (P. X.) BM is greater than AB.

Quod erat hoc loco intentum.

Since therefore, assuming in AD produced the interval AF double the interval AD, the perpendicular FM let fall on the transversal AP cuts off from it more than double what is cut off by the perpendicular let fall from the point D: [16] more quickly by far in this hypothesis of obtuse angle, than in the preceding hypothesis of right angle, we attain to an interval so great, that from it the perpendicular let fall cuts off a base greater than the designated AP however great.

But this, as in the preceding proposition, could not happen, unless after the meeting of the produced AD with PL in some point; and indeed at a finite, or terminated distance.

Quod erat etc.

PROPOSITIO XIII.

Si recta XA (quantaelibet designatae longitudinis) incidens in duas rectas AD, XL, efficiat cum eisdem ad easdem partes (fig. 11.) angulos internos XAD, AXL minores duobus rectis: dico, illas duas (etiamsi neuter illorum angulorum sit rectus) tandem in aliquo puncto ad partes illorum angulorum invicem coituras, et quidem ad finitam, seu terminatam distantiam, dum consistat alterutra hypothesis aut anguli recti, aut anguli obtusi.

Demonstratur. Nam unus praedictorum angulorum, ut puta AXL, erit acutus. Itaque ex apice alterius anguli demittatur ad XL perpendicularis AP, quae utique (propter decimamseptimam primi) cadet ad partes anguli acuti AXL. Quoniam igitur in triangulo APX, rectangulo in P, duo simul anguli acuti PAX, PXA, minores non sunt (ex nona hujus) uno recto, in utraque hypothesi aut anguli recti, aut anguli obtusi; si duo isti anguli auferantur in summa angulorum propositorum jam reliquus angulus PAD minor erit recto. Itaque erimus in casu duarum praecedentium Propositionum, dum scilicet alterutra hypothesis consistat aut anguli recti, aut anguli obtusi. Quare (ex eisdem) rectae AD, et PL, sive XL, in aliquo puncto finitae, seu terminatae distantiae ad notas [17] partes concurrent, tam sub una, quam sub altera praedictarum hypothesium. Quod erat demonstrandum.

PROPOSITION XIII. ✲

If the straight XA (of designated length however great) meeting two straights AD, XL, makes with them toward the same parts (fig. 11) internal angles XAD, AXL less than two right angles: I say, these two (even if neither of those angles be a right angle) at length will mutually meet in some point on the side toward those angles, and indeed at a finite, or terminated distance, if either hypothesis holds, of right angle or of obtuse angle.

PROOF. For one of the said angles, as AXL suppose, will be acute.

Accordingly from the vertex of the other angle is dropped the perpendicular AP on XL, which certainly

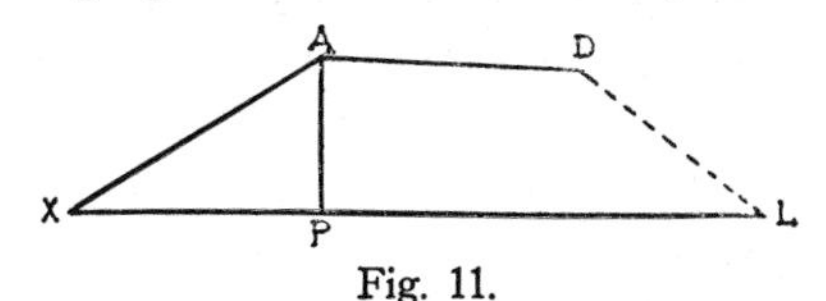

Fig. 11.

(because of Eu. I. 17) falls on the side of the acute angle AXL. Since therefore in the triangle APX, right-angled at P, the two acute angles PAX, PXA, together are not less (P. IX.) than a right angle, in either hypothesis, of right angle, or of obtuse angle; if these two angles are taken away from the sum of the given angles the then remaining angle PAD will be less than a right angle. Consequently we will be in the case of the two preceding propositions, since it is obvious that one or the other hypothesis holds, either of right angle, or of obtuse angle.

Wherefore the straights AD, and PL, or XL, meet in some point at a finite, or terminated distance on the side noted, [17] as well under the one as under the other mentioned hypothesis.

Quod erat demonstrandum.

SCHOLION I.

Ubi observare licet notabile discrimen ab hypothesi anguli acuti. Nam in ista demonstrari nequiret generalis hujusmodi rectarum concursus, quoties recta aliqua in duas incidens, duos ad easdem partes efficiat internos angulos duobus rectis minores; nequiret, inquam, directe demonstrari, etiamsi in eadem hypothesi admitteretur praedictus generalis concursus, quoties unus duorum angulorum est rectus. Quamvis enim recta AD perpendicularis et ipsa foret rectae AP; quo casu nequiret certe, propter 17. primi, concurrere cum altera perpendiculari PL; nihilominus duo simul anguli DAX, PXA, minores forent duobus rectis, juxta hypothesim praedictam, cum in ea duo simul anguli PAX, PXA minores sint (ex nona hujus) uno recto. Id autem observasse operae pretium fuit.

Qualiter vero ex eo solo admisso generali concursu, dum unus angulorum est rectus, et quidem sub assignata quantumlibet parva incidente, destrui possit hypothesis anguli acuti; docebimus post tres sequentes Propositiones.

SCHOLION II.

In tribus ante jactis theorematis studiose apposui illam conditionem, quod recta incidens AP, sive XA, intelligatur esse *quantaelibet designatae longitudinis*. Si enim, citra omnem rectae incidentis determinatam mensuram, praecise agatur de exhibendo, ac demonstrando duarum rectarum concursu in apicem cujusdam trianguli, cujus [18] anguli ad basim sint dati (minores utique duobus rectis) ut puta unus rectus, et alter duobus tantum

SCHOLION I.

Here may be observed a notable difference from the hypothesis of acute angle.

For in this the general concurrence of such straights cannot be demonstrated, as often as any straight falling upon two, makes two internal angles toward the same parts less than two right angles; cannot, I say, be directly demonstrated, even if in this hypothesis the aforesaid general concurrence be admitted, as often as one of the two angles is right.

For although the straight AD be perpendicular even to the straight AP; in which case it certainly could not concur with another perpendicular PL (Eu. I. 17); nevertheless the two angles together DAX, PXA, could be less than two right angles, in accordance with the aforesaid hypothesis, since in it the two angles together PAX, PXA may be less (P. IX.) than one right angle. ✡ But it was worth while to have observed this.

But how, solely from the general admission of concurrence when one of the angles is right, and with an assigned incident however small, the hypothesis of acute angle can be demolished; this we shall show after the next three propositions.

SCHOLION II. ✡

In the three preceding theorems I have studiously set down this condition, that the cutting straight AP, or XA, is understood to be of a *designated length as great as you choose.*

For if, without any determinate extent of the cutting straight, it be discussed precisely concerning the exhibiting and demonstrating of the concurrence of two straights at the apex of a certain triangle, whose [18] angles at the base are given (less indeed than two right angles) as,

gradibus, vel, ut libet, minus deficiens a recto; quis est tam expers Geometriae, quin statim rem ipsam demonstrative exhibeat? Nam supponatur (fig. 12.) datus quilibet angulus BAP, ut puta 88. graduum. Si ergo ex quolibet puncto B ipsius AB, demittatur ad subjectam AP (juxta duodecimam primi) perpendicularis BP, constat enim vero in eo triangulo ABP exhibitum fore demonstrative concursum optatum in eo puncto B.

Quod si alter angulus ad basim postuletur et ipse minor recto, ut puta 84. graduum, quem nempe exhibeat datus angulus K: tunc (juxta 23. primi) efficere poteris versus partes rectae AB aequalem angulum APD, occurrente PD ipsi AB in quodam ejus intermedio puncto D. Quare habebitur rursum demonstrative concursus optatus in eo puncto D.

Tandem vero: si alter angulus postuletur obtusus, sed minor tamen 92. gradibus, ne cum alio dato angulo BAP compleantur duo recti: exhibitus hic sit in quodam angulo R 91. graduum. Ostendendum est, unum aliquod esse punctum X *in* ipsa AP, ad quod juncta BX efficiat angulum BXA aequalem dato angulo R 91. graduum; adeo ut propterea sub quadam recta incidente AX habeatur concursus optatus in praedicto puncto B. Sic autem proceditur. Quandoquidem (protracta PA usque in aliquod punctum H) angulus externus BAH et est (propter decimamtertiam primi) 92. graduum, cum angulus interior BAP positus sit 88. graduum; ac rursum, propter

suppose, one right, and the other less than a right by as much as two degrees, or, if you please, by less; who is so devoid of geometry that he could not immediately show the thing itself demonstratively?

For suppose (fig. 12) given any angle BAP, as, say, 88 degrees. If therefore from any point B of this AB, is let fall on the base AP (Eu. I. 12) the perpendicular BP, it holds certainly that in this triangle ABP would be exhibited demonstratively the desired concurrence at this point B. ☼

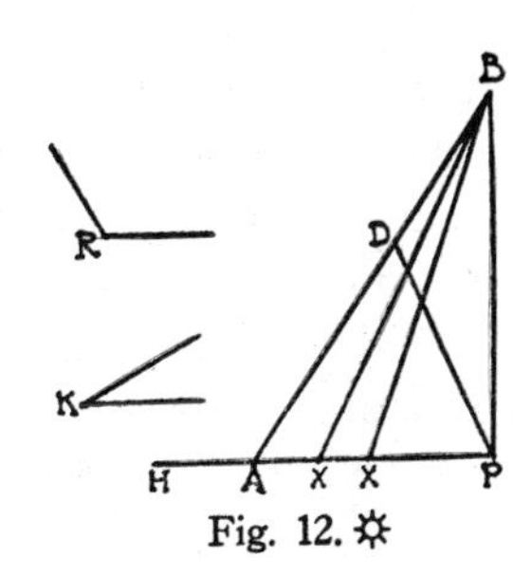

Fig. 12. ☼

But if the other angle at the base is postulated, and is less than a right, as, suppose, 84 degrees, which indeed ☼ the given angle K represents: then (Eu. I. 23) one would be able to make toward the parts of the straight AB an equal angle APD, PD meeting this AB in D, some intermediate point of it. Wherefore the desired concourse is again obtained demonstratively in this point D.

But finally: if the other angle is postulated obtuse, but yet less than 92 degrees, lest with the other given angle BAP it should make up two rights: this may be represented in a certain angle R of 91 degrees. It is to be shown, that there is some point X of this AP, to which the join BX makes an angle BXA equal to the given angle R of 91 degrees: so that therefore under a certain cutting straight AX the desired meeting in the point B may be obtained.

Now we may proceed thus.

PA being produced to any point H, since the external angle BAH is (Eu. I. 13) 92 degrees, because the interior angle BAP is by hypothesis 88 degrees; and again (Eu.

decimamsextam primi, major est non solum angulo recto BPA, verum etiam quibusvis eodem titulo obtusis angulis BXA, sumpto puncto X ubilibet intra ipsam PA, et quidem, propter eandem decimamsextam primi, semper majoribus, dum punctum X assumitur propius puncto A: consequens pla-[19]ne est, ut inter istos angulos, unum 90. graduum in puncto P, et alterum 92. graduum in puncto A, unus reperiatur angulus BXA, qui sit 91. graduum, nimirum aequalis dato angulo R.

Nihilominus, omissa postrema hac observatione circa angulum obtusum, cavere diligentissime oportet, in eo positam esse difficultatem illius pronunciati Euclidaei, quod velit occursum duarum rectarum; in illam utique partem, ad quam cum recta incidente duos angulos efficiant duobus rectis minores; atque ita quidem praedictum occursum velit, *quantaecunque longitudinis sit incidens assignata.* Caeterum enim (ut jam monui in praecedente Scholio) demonstrabo generalem istum occursum ex solo admisso occursu ejusmodi, dum unus angulorum sit rectus; et quidem, etiamsi admisso non pro qualibet assignabili finita incidente, sed solum admisso intra limites cujusdam assignatae parvissimae incidentis.

PROPOSITIO XIV.

Hypothesis anguli obtusi est absolute falsa, quia se ipsam destruit.

Demonstratur. Ex hypothesi anguli obtusi, assumpta ut vera, jam elicuimus veritatem illius Pronunciati Euclidaei; quod duae rectae sibi invicem in aliquo puncto ad eas partes occursurae sint, ad quas recta quaedam, easdem secans, duos qualescunque effecerit internos angulos, duobus rectis minores. Stante autem hoc Pronunciato, cui

I. 16) is greater not alone than the right angle BPA but also, for the same reason, than any obtuse angle BXA, the point X being assumed wherever you choose within this PA, and indeed always greater as the point X is assumed nearer to the point A (Eu. I. 16): it is an evident consequence, [19] that between those angles, one of 90 degrees at the point P, and the other of 92 degrees at the point A, one angle BXA is found, which is 91 degrees, truly equal to the given angle R. ✲

None the less, omitting here the last observation about the obtuse angle, it is necessary most diligently to take care that the difficulty of this assumption of Euclid be fixed in this, that it asserts the meeting of two straights; in particular in that part toward which they make with the cutting straight two angles together less than two right angles; and assuredly that it asserts the aforesaid meeting thus, *of whatever length be the assigned transversal.*

However (as I have already mentioned in the preceding scholion) I shall demonstrate the general meeting ✲ solely from the admitted meeting of this sort when one of the angles is right; and indeed even if it be admitted not for any assignable finite transversal, but alone admitted within the limits of any assigned very small transversal.

PROPOSITION XIV.

The hypothesis of obtuse angle is absolutely false, because it destroys itself.

Proof. From the hypothesis of obtuse angle, assumed as true, we have now deduced the truth of Euclid's postulate: that two straights will meet each other in some point toward those parts, toward which a certain straight, cutting them, makes two internal angles, of whatever kind, less than two right angles.

innititur Euclides post vigesimamoctavam sui Libri primi, manifestum est omnibus Geometris, solam hypothesim anguli recti esse veram, nec ullum relinqui locum hypothesi anguli obtusi. Igitur hypothesis anguli obtusi est absolute falsa, quia se ipsam destruit. Quod erat demonstrandum. [20]

Aliter, ac magis immediate. Quandoquidem ex hypothesi anguli obtusi demonstravimus (in nona hujus) duos (fig. 11.) acutos angulos trianguli APX, rectanguli in P, majores esse uno recto; constat talem assumi posse acutum angulum PAD, qui simul cum praedictis duobus acutis angulis duos rectos efficiat. Tunc autem recta AD deberet (ex praecedente, juxta hypothesim anguli obtusi) aliquando concurrere cum ipsa PL, sive XL, respectu habito ad secantem, sive incidentem AP; quod est manifestum absurdum contra decimamseptimam primi, si respicias ad secantem, sive incidentem AX.

PROPOSITIO XV.

Ex quolibet triangulo ABC, cujus tres simul anguli (fig. 13.) aequales sint, aut majores, aut minores duobus rectis, stabilitur respective hypothesis aut anguli recti, aut anguli obtusi, aut anguli acuti.

Demonstratur. Nam duo saltem illius trianguli anguli, ut puta ad puncta A, et C, acuti erunt, propter decimamseptimam primi. Quare perpendicularis, ex apice reliqui anguli B ad ipsam AC demissa, secabit ipsam AC (propter eandem decimamseptimam primi) in aliquo puncto intermedio D. Si ergo tres anguli ipsius trianguli ABC supponantur aequales duobus rectis, constat aequales

But this assumption holding good, on which Euclid supports himself after I. 28, it is manifest to all geometers that the hypothesis of right angle alone is true, nor any place left for the hypothesis of obtuse angle. Therefore the hypothesis of obtuse angle is absolutely false, because it destroys itself.

Quod erat demonstrandum. [20]

Otherwise, and more immediately. Since from the hypothesis of obtuse angle we have proved (P. IX.) that two (fig. 11) acute angles of the triangle APX, right-angled at P, are greater than one right angle; it follows that an acute angle PAD may be assumed such, that together with the aforesaid two acute angles it makes up two right angles. But then the straight AD must (by the preceding proposition, joined to the hypothesis of obtuse angle) at length meet with this PL, or XL, regard being had to the secant, or incident AP; which is manifestly absurd (against Eu. I. 17) if we regard the secant, or incident AX.

PROPOSITION XV. ✲

By any triangle ABC, of which the three angles (fig. 13) are equal to, or greater, or less than two right angles, is established respectively the hypothesis of right angle, or obtuse angle, or acute angle.

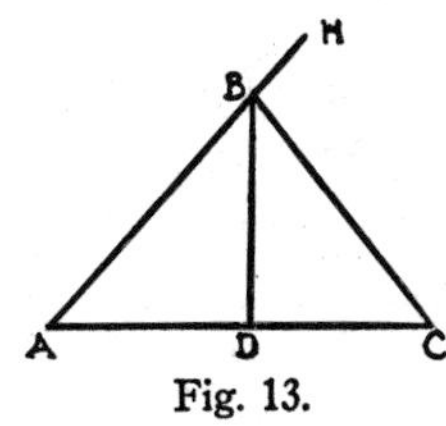

Fig. 13.

Proof. For anyhow two angles of this triangle, as suppose at the points A and C, will be acute (Eu. I. 17). Wherefore the perpendicular, let fall from the apex of the remaining angle B upon AC, will cut AC (Eu. I. 17) in some intermediate point D.

If therefore the three angles of this triangle ABC are supposed equal to two right angles, it follows that

fore quatuor rectis omnes simul angulos triangulorum ADB, CDB, propter duos additos rectos angulos ad punctum D. Hoc stante: neutrius modo dictorum triangulorum, ut puta ADB, tres simul anguli minores erunt, aut majores duobus rectis; nam sic viceversa alterius trianguli tres simul anguli majores forent, aut minores duobus rectis. Quare (ex nona hujus) ab uno quidem triangulo stabiliretur hypothesis anguli acuti, et ab altero hypothesis anguli [21] obtusi; quod repugnat sextae, et septimae hujus. Igitur tres simul anguli utriusque praedictorum triangulorum aequales erunt duobus rectis; ac propterea (ex nona hujus) stabilietur hypothesis anguli recti. Quod erat primo loco demonstrandum.

Sin autem tres anguli propositi trianguli ABC ponantur majores duobus rectis; jam duorum triangulorum ADB, CDB omnes simul anguli majores erunt quatuor rectis, propter duos additos rectos angulos ad punctum D. Hoc stante: neutrius modo dictorum triangulorum tres simul anguli aequales praecise erunt, aut minores duobus rectis; nam sic viceversa alterius trianguli tres simul anguli majores forent duobus rectis. Quare (ex nona hujus) ab uno quidem triangulo stabiliretur hypothesis aut anguli recti, aut anguli acuti, et ab altero hypothesis anguli obtusi, quod repugnat quintae, sextae, et septimae hujus. Igitur tres simul anguli utriusque praedictorum triangulorum majores erunt duobus rectis; ac propterea (ex nona hujus) stabilietur hypothesis anguli obtusi. Quod erat secundo loco demonstrandum.

all the angles of the triangles ADB, CDB will be together equal to four right angles, because of the two additional right angles at the point D. This holding good, now of neither of the said triangles, as suppose ADB, will the three angles together be less, or greater than two right angles; for thus *vice versa* the three angles together of the other triangle would be greater, or less than two right angles. Wherefore (P. IX.) from one triangle would indeed be established the hypothesis of acute angle, and from the other the hypothesis of obtuse angle; [21] which is contrary to P. VI. and P. VII.

Therefore the three angles together of either of the aforesaid triangles will be equal to two right angles; and thereby (P. IX.) is established the hypothesis of right angle.

Quod erat primo loco demonstrandum.

But if however the three angles of the proposed triangle ABC are taken greater than two right angles; now of the two triangles ADB, CDB all the angles together will be greater than four right angles, because of the two additional right angles at the point D.

This holding good: now of neither of the said triangles will the three angles together be precisely equal to, or less than two right angles: for thus *vice versa* the three angles of the other triangle would be together greater than two right angles. Wherefore (P. IX.) from one triangle indeed would be established the hypothesis either of right angle or of acute angle, and from the other the hypothesis of obtuse angle, which is contrary to Propp. V., VI., and VII.

Therefore the three angles together of either of the aforesaid triangles will be greater than two right angles; and therefore is established the hypothesis of obtuse angle.

Quod erat secundo loco demonstrandum.

Tandem vero. Si tres anguli propositi trianguli ABC ponantur minores duobus rectis, jam duorum triangulorum ADB, CDB, omnes simul anguli minores erunt quatuor rectis, propter duos additos rectos angulos ad punctum D. Hoc stante: neutrius modo dictorum triangulorum tres simul anguli aequales erunt, aut majores duobus rectis; nam sic viceversa alterius trianguli tres simul anguli minores forent duobus rectis. Quare (ex nona hujus) ab uno quidem triangulo stabiliretur hypothesis aut anguli recti, aut anguli obtusi, et ab altero hypothesis anguli acuti; quod repugnat quintae, sextae, et septimae hujus. Igitur tres simul anguli utriusque praedictorum triangulorum minores erunt duobus rectis; ac propterea (ex [22] nona hujus) stabilietur hypothesis anguli acuti. Quod erat tertio loco demonstrandum.

Itaque ex quolibet triangulo ABC, cujus tres simul anguli aequales sint, aut majores, aut minores duobus rectis, stabilitur respective hypothesis aut anguli recti, aut anguli obtusi, aut anguli acuti. Quod erat propositum.

COROLLARIUM.

Hinc; protracto uno quolibet cujusvis propositi trianguli latere, ut puta AB in H; erit (ex 13. primi) externus angulus HBC aut aequalis, aut minor, aut major reliquis simul internis, et oppositis angulis ad puncta A, et C, prout vera fuerit hypothesis aut anguli recti, aut anguli obtusi, aut anguli acuti. Et vicissim.

But finally. If the three angles of the proposed triangle ABC are taken less than two right angles, now of the two triangles ADB, CDB, all the angles together will be less than four right angles, because of the two additional right angles at the point D.

This holding good: now of neither of the said triangles will the three angles together be equal to, or greater than two right angles; for thus *vice versa* of the other triangle the three angles together would be less than two right angles.

Wherefore (P. IX.) from one triangle indeed would be established the hypothesis either of right angle or of obtuse angle, and from the other the hypothesis of acute angle; which is contrary to Propp. V., VI., and VII.

Therefore the three angles together of either of the aforesaid triangles will be less than two right angles; and therefore (P. IX.) [22] is established the hypothesis of acute angle.

Quod erat tertio loco demonstrandum.

Accordingly by any triangle ABC, of which the three angles are together equal to, or greater, or less than two right angles, is established respectively the hypothesis of right angle, or obtuse angle, or acute angle.

Quod erat propositum.

COROLLARY.

Hence, any one side of any proposed triangle being produced, as suppose AB to H; the external angle HBC will be (Eu. I. 13) equal to, or less, or greater than the remaining internal and opposite angles together at the points A, and C, according as is true the hypothesis of right angle, or obtuse angle, or acute angle. And inversely.

PROPOSITIO XVI.

Ex quolibet quadrilatero ABCD, cujus quatuor simul anguli aequales sint, aut majores, aut minores quatuor rectis, stabilitur respective hypothesis aut anguli recti, aut anguli obtusi, aut anguli acuti.

Demonstratur. Jungatur AC. Non erunt (fig. 14.) tres simul anguli trianguli ABC aequales, aut majores, aut minores duobus rectis, quin tres simul anguli trianguli ADC sint ipsi etiam respective aequales, aut majores, aut minores duobus rectis; ne scilicet (ex praecedente) ab uno illorum triangulorum stabiliatur una hypothesis, et ab altero altera, contra quintam, sextam, et septimam hujus. Hoc stante: Si quatuor simul anguli propositi quadrilateri aequales sint quatuor rectis, constat utriusque modo dictorum triangulorum tres simul angulos aequales fore duobus rectis, atque ideo (ex praecedente) stabili-[23]tum iri hypothesim anguli recti.

Sin vero ejusdem quadrilateri quatuor simul anguli majores sint, aut minores quatuor rectis, debebunt similiter illorum triangulorum tres simul anguli respective esse aut una majores, aut una minores duobus rectis. Quare ab illis triangulis stabilietur respective (ex praecedente) aut hypothesis anguli obtusi, aut hypothesis anguli acuti.

Itaque ex quolibet quadrilatero, cujus quatuor simul anguli aequales sint, aut majores, aut minores quatuor

PROPOSITION XVI.

By any quadrilateral ABCD, of which the four angles together are equal to, or greater, or less than four right angles, is established respectively the hypothesis of right angle, or obtuse angle, or acute angle.

PROOF. Join AC. The three angles of the triangle ABC (fig. 14) will not be together equal to, or greater, or less than two right angles, without the three angles of the triangle ADC being themselves also together respectively equal to, or greater, or less than two right angles, lest obviously (by the preceding) from one of those triangles be established one hypothesis, and another from the other, against the fifth, sixth, and seventh propositions of this work.

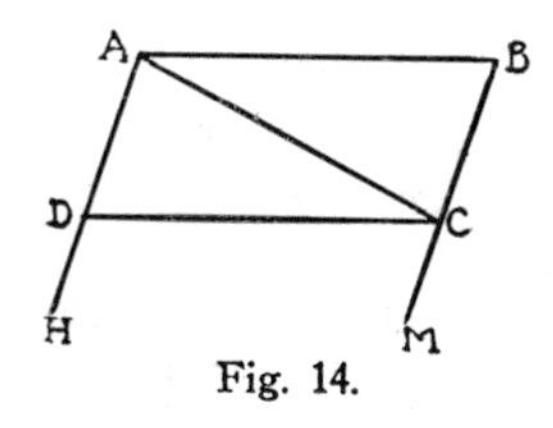

Fig. 14.

This holding good: If the four angles together of the premised quadrilateral are equal to four right angles, it follows that the three angles together of either of the just mentioned triangles will be equal to two right angles, and therefore (from the preceding) [23] the hypothesis of right angle will be established.

But if indeed the four angles of this quadrilateral be together greater, or less than four right angles, similarly the three angles together of those triangles should be respectively either at the same time greater, or at the same time less than two right angles. Wherefore from these triangles would be established respectively (from the preceding) either the hypothesis of obtuse angle, or the hypothesis of acute angle.

Therefore by any quadrilateral, of which the four angles together are equal to, or greater, or less than four

rectis, stabilitur respective hypothesis aut anguli recti, aut anguli obtusi, aut anguli acuti. Quod erat demonstrandum.

COROLLARIUM.

Hinc: protractis versus easdem partes duobus quibusvis propositi quadrilateri contrapositis lateribus, ut puta AD in H, et BC in M; erunt (ex 13. primi) duo simul externi anguli HDC, MCD aut aequales, aut minores, aut majores duobus simul internis, et oppositis angulis ad puncta A, et B, prout vera fuerit hypothesis aut anguli recti, aut anguli obtusi, aut anguli acuti.

PROPOSITIO XVII.

Si uni, ut libet, cuidam parvae rectae AB insistat (fig. 15.) ad rectos angulos recta AH: Dico subsistere non posse in hypothesi anguli acuti, ut quaevis BD, efficiens cum AB quemlibet angulum acutum versus partes ipsius AH, occursura tandem sit ad finitam, seu terminatam distantiam ipsi AH productae.

Demonstratur. Jungatur HB. Erit (ex 17. primi) acutus angulus ABH, propter angulum rectum ad punctum A. Jam (ex 23. primi) ducatur quaedam HD ver-[24]sus partes puncti B, quae non secans angulum AHB efficiat cum ipsa HB angulum acutum aequalem ipsi acuto ABH. Deinde ex puncto B demittatur ad HD perpendicularis BD, quae cadet ad partes praedicti anguli acuti ad punctum H. Quoniam igitur latus HB opponitur in triangulo HDB angulo recto in D, atque item in triangulo

right angles, is established respectively the hypothesis of right angle, or obtuse angle, or acute angle.

Quod erat demonstrandum.

COROLLARY.

Hence, any two opposite sides of the premised quadrilateral being produced toward the same parts, as suppose AD to H, and BC to M; the two external angles HDC, MCD will be (Eu. I. 13) either equal to, or less, or greater than the two internal and opposite angles together at the points A, and B, according as is true the hypothesis of right angle, or obtuse angle, or acute angle.

PROPOSITION XVII.

If the straight AH stands (fig. 15) at right angles to any certain arbitrarily small straight AB: I say that in the hypothesis of acute angle it cannot hold good, that every straight BD, making with AB toward the parts of this AH any acute angle you choose, will at length meet this AH produced at a finite, or terminated distance.

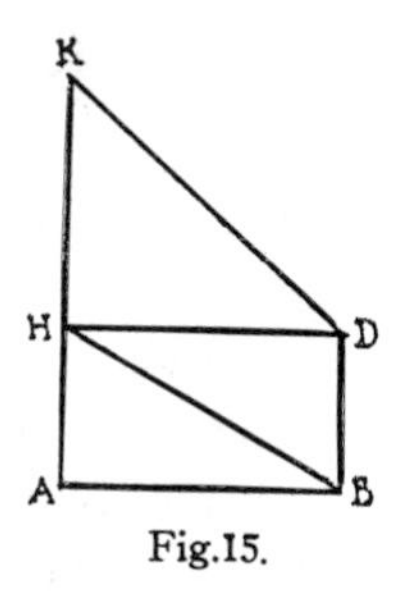

Fig.15.

Proof. Join HB. The angle ABH will be acute (Eu. I. 17) because of the right angle at the point A. Now draw (Eu. I. 23) HD toward [24] the parts of the point B, which not cutting the angle AHB makes with this HB an acute angle equal to this acute angle ABH. Then from the point B is let fall to HD the perpendicular BD, which will fall toward the parts of the aforesaid acute angle at the point H.

Since therefore the side HB is opposite in the triangle HDB to the right angle at D, and likewise in the triangle

BAH angulo recto in A; ac rursum in duobus illis triangulis adjacent eidem lateri HB aequales anguli, qui sunt in priore quidem triangulo angulus BHD, et in posteriore angulus HBA; erit etiam (ex 26. primi) reliquus angulus HBD in priore triangulo aequalis reliquo angulo BHA in posteriore triangulo. Quare integer angulus DBA aequalis erit integro angulo AHD.

Jam vero: non erit uterque praedictorum aequalium angulorum obtusus, ne incidamus (ex praecedente) in unum casum jam reprobatae hypothesis anguli obtusi. Sed neque erit rectus, ne incidamus (ex eadem praecedente) in unum casum pro hypothesi anguli recti, qui nullum (ex 5. hujus) relinqueret locum hypothesi anguli acuti. Uterque igitur illorum angulorum erit acutus. Hoc stante: Quod recta BD protracta occurrere nequeat in quodam puncto K ipsi AH ad easdem partes productae, ex eo demonstratur; quia in triangulo KDH, praeter angulum rectum in D, adesset angulus obtusus in H, cum angulus AHD, in praedicta hypothesi anguli acuti, demonstratus sit acutus. Hoc autem absurdum est, contra 17. primi. Non ergo subsistere potest in ea hypothesi, ut quaevis BD, efficiens cum una, ut libet parva recta AB, quemlibet angulum acutum versus partes ipsius AH, occursura tandem sit ad finitam, seu terminatam distantiam, ipsi AH productae. Quod erat demonstrandum.

Aliter idem, ac facilius. Insistant uni cuidam quantumlibet parvae rectae AB (fig. 16.) duae perpendiculares [25] AK, BM. Demittatur ad AK ex aliquo puncto M ipsius BM perpendicularis MH, jungaturque BH. Constat acutum fore angulum BHM. Est etiam (ex praece-

BAH to the right angle at A; and again in those two triangles equal angles are adjacent to this side HB, which are in the first triangle indeed the angle BHD, and in the latter the angle HBA; also (Eu. I. 26) the remaining angle HBD in the former triangle will be equal to the remaining angle BHA in the latter triangle. Wherefore the entire angle DBA will be equal to the entire angle AHD.

Now however, neither of the aforesaid equal angles will be obtuse, lest we meet (from the preceding proposition) a case of the now rejected hypothesis of obtuse angle.

Nor will either be right, lest we meet (from the same preceding) a case of the hypothesis of right angle, which (P. V.) will leave no place for the hypothesis of acute angle. Therefore each one of those angles will be acute. This being the case: that the straight BD produced cannot meet in a certain point K this AH produced toward the same parts, is demonstrated thus; because in the triangle KDH, besides the right angle at D, is present the obtuse angle at H, since the angle AHD in the aforesaid hypothesis of acute angle is proved acute. But this is absurd, against Eu. I. 17.

Therefore it cannot hold good in this hypothesis, that any BD, making with an arbitrarily small straight AB any acute angle toward the parts of this AH, will at length at a finite, or terminated distance, meet this AH produced.

Quod erat demonstrandum.

The same otherwise and more easily.

Two perpendiculars AK, BM stand on a certain straight AB, as small as you choose (fig. 16). [25] From any point M of this BM let fall to AK the perpendicular MH, and join BH. It follows that the angle BHM will

dente) acutus angulus BMH, in hypothesi anguli acuti. Ergo perpendicularis BDX, ex puncto B ad ipsam HM demissa, secabit (ex 17. primi) eam HM in quodam puncto intermedio D. Ergo angulus XBA erit acutus. Constat autem (ex eadem 17. primi) non posse invicem concurrere (saltem ad finitam, seu terminatam distantiam) duas illas utcunque productas AHK, BDX, propter angulos rectos in punctis H, et D. Itaque nequit subsistere in hypothesi anguli acuti, ut quaevis BD, efficiens cum una, ut libet, parva recta AB, quemlibet angulum acutum versus partes ipsius AH, eidem AB perpendicularis, occursura tandem sit (ad finitam, seu terminatam distantiam) ipsi AH productae. Quod erat propositum.

SCHOLION I.

Atque id est, quod spopondi in Scholiis post XIII. hujus, nimirum destructum iri hypothesim anguli acuti (quae sola obesse jam potest generali illi Pronunciato Euclidaeo) ex solo admisso generali duarum rectarum concursu ad eas partes, versus quas recta quaepiam, quantumlibet parva, in easdem incidens, duos efficiat internos angulos minores duobus rectis; atque ita quidem, etiamsi alteruter illorum angulorum supponi debeat rectus.

SCHOLION II.

Sed rursum meliore loco, post XXVII. hujus, ostendam destructum pariter iri hypothesim anguli acuti, dum unus aliquis tenuissimus, ut libet, angulus acutus desi-[26]gnari

be acute. In the hypothesis of acute angle, the angle BMH is also (from the preceding proposition) acute. Therefore the perpendicular BDX, let fall from the point B to this HM, will cut (by Eu. I. 17) this HM in some intermediate point D. Therefore the angle XBA will be acute.

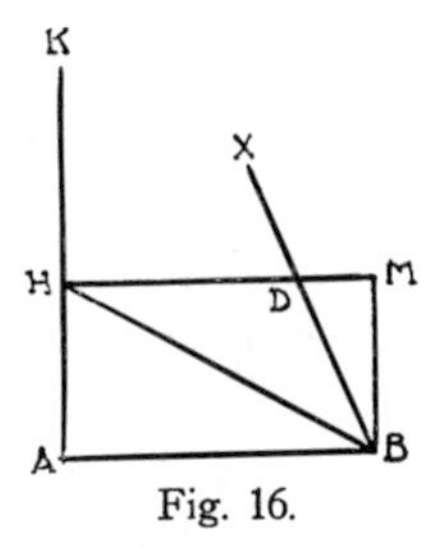

Fig. 16.

But it follows (Eu. I. 17) that those two straights AHK, BDX however produced cannot meet (anyhow at a finite or terminated distance) on account of the right angles at the points H and D. Therefore in the hypothesis of acute angle it cannot hold good, that any BD, making with a straight AB, however small, any acute angle toward the parts of this AH, perpendicular to this same AB, will at length meet (at a finite or terminated distance) this AH produced.

Quod erat propositum.

SCHOLION I.

And this is what I promised in the scholia after P. XIII., that the hypothesis of acute angle (which alone is able now to stand against that general Euclidean assumption) will certainly be destroyed by the sole admission of a universal meeting of two straights toward those parts toward which any straight, as small as you choose, meeting them, makes two internal angles less than two right angles; and just so, even if either of those angles is to be supposed right.

SCHOLION II.

But again in a better place, after P. XXVII., I shall show that the hypothesis of acute angle will be equally destroyed, provided that any one acute angle as small as

possit; sub quo, si recta quaepiam in alteram incidat, debeat haec producta (ad finitam, seu terminatam distantiam) aliquando occurrere cuivis ad quantamlibet finitam distantiam excitatae super ea incidente perpendiculari.

PROPOSITIO XVIII.

Ex quolibet triangulo ABC, cujus angulus (*fig.* 17.) *ad punctum B in uno quovis semicirculo existat, cujus diameter AC, stabilitur hypothesis aut anguli recti, aut anguli obtusi, aut anguli acuti, prout nempe angulus ad punctum B fuerit aut rectus, aut obtusus, aut acutus.*

Demonstratur. Ex centro D jungatur DB. Erunt (ex quinta primi) aequales anguli ad basim AB, atque item ad basim BC, in triangulis ADB, CDB. Quare, in triangulo ABC, duo simul anguli ad basim AC aequales erunt toti angulo ABC. Igitur tres simul anguli trianguli ABC aequales erunt, aut majores, aut minores duobus rectis, prout angulus ad punctum B fuerit aut rectus, aut obtusus, aut acutus. Itaque ex quolibet triangulo ABC, cujus angulus ad punctum B in uno quovis semicirculo existat, cujus diameter AC, stabilitur (ex 15. hujus) hypothesis aut anguli recti, aut anguli obtusi, aut anguli acuti, prout nempe angulus ad punctum B fuerit aut rectus, aut obtusus, aut acutus. Quod erat etc.

PROPOSITIO XIX.

Esto quodvis triangulum AHD (*fig.* 18.) *rectangulum in H. Tum in AD continuata sumatur portio DC aequalis ipsi AD; demittaturque ad AH productam*

you choose can be designated,[26] under which if any straight line meets another, this produced must (at a finite or terminated distance) finally meet any perpendicular erected upon this incident straight at whatever finite distance.

PROPOSITION XVIII.

From any triangle ABC, of which (fig. 17) the angle at the point B is inscribed in any semicircle of diameter AC, is established the hypothesis of right angle, or obtuse angle, or acute angle, according as indeed the angle at the point B is right, or obtuse, or acute.

PROOF. From the center D join DB. The angles at the base AB will be (Eu. I. 5) equal, and likewise at the base BC, in the triangles ADB, CDB. Wherefore, in the triangle ABC the two angles at the base AC will be together equal to the whole angle ABC. Therefore the three angles of the triangle ABC will be together equal to, or greater, or less than two right angles, according as the angle at the point B is right, or obtuse, or acute.

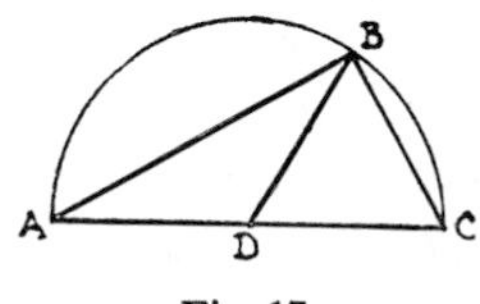

Fig. 17.

Therefore from any triangle ABC, of which the angle at the point B is inscribed in any semicircle of diameter AC, is established (P. XV.) the hypothesis of right angle, or obtuse angle, or acute angle, according as indeed the angle at the point B is right, or obtuse, or acute.

Quod erat demonstrandum.

PROPOSITION XIX.

Let there be any triangle AHD (fig. 18) right-angled at H. Then in AD produced the portion DC is assumed equal to this AD; and the perpendicular CB

perpendicularis CB. Dico stabilitum hinc iri hypothesim aut anguli recti, aut anguli obtusi, aut anguli acuti, prout portio HB aequalis fue-[27]*rit, aut major, aut minor ipsa AH.*

Demonstratur. Nam juncta DB erit (ex 4. primi, et ex 10. hujus) aut aequalis, aut major, aut minor ipsa AD, sive DC, prout illa portio HB aequalis fuerit, aut major, aut minor ipsa AH.

Et primo quidem sit HB aequalis ipsi AH, ita ut propterea juncta DB aequalis sit ipsi AD, sive DC. Constat circumferentiam circuli, qui centro D, et intervallo DB describatur, transituram per puncta A, et C. Igitur angulus ABC, qui ponitur rectus, existet in eo semicirculo, cujus diameter AC. Quare (ex praecedente) stabilietur hypothesis anguli recti. Quod erat primo loco demonstrandum.

Sit secundo HB major ipsa AH, ita ut propterea juncta DB major sit ipsa AD, sive DC. Constat circumferentiam circuli, qui centro D, et intervallo DA, sive DC, describatur, occursuram ipsi DB in aliquo puncto intermedio K. Igitur, junctis AK, et CK, erit angulus AKC obtusus, quia major (ex 21. primi) angulo ABC, qui ponitur rectus. Quare (ex praecedente) stabilietur hypothesis anguli obtusi. Quod erat secundo loco demonstrandum.

Sit tertio HB minor ipsa AH, ita ut propterea juncta DB minor sit ipsa AD, sive DC. Constat circumferentiam circuli, qui centro D, et intervallo DA, sive DC describatur, occursuram in aliquo puncto M ipsius DB

is let fall to AH produced. I say hence will be established the hypothesis of right angle, or obtuse angle, or acute angle, according as the portion HB is equal to, [27] *or greater, or less than AH.*

PROOF. For the join DB will be (Eu. I. 4, and P. X. of this) either equal to, or greater, or less than AD, or DC, according as the portion HB is equal to, or greater, or less than AH.

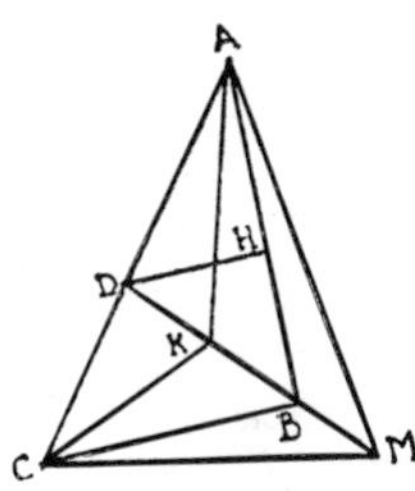

Fig. 18.

And first indeed let HB be equal to AH, so that therefore the join DB may be equal to AD, or DC. It follows that the circumference of the circle, which is described with the center D and radius DB, will go through the points A and C. Therefore the angle ABC, which is assumed right, is in this semicircle, whose diameter is AC. Wherefore (from the preceding proposition) is established the hypothesis of right angle.

Quod erat primo loco demonstrandum.

Secondly let BH be greater than AH, so that therefore the join DB is greater than AD, or DC. It follows that the circumference of the circle, which is described with center D, and radius DA, or DC, will meet DB in some intermediate point K. Therefore, AK, and CK being joined, the angle AKC will be obtuse, because greater (Eu. I. 21) than the angle ABC, which is assumed right. Wherefore (from the preceding proposition) is established the hypothesis of obtuse angle.

Quod erat secundo loco demonstrandum.

Thirdly let BH be less than AH, so that therefore the join DB is less than AD, or DC. It follows that the circumference of the circle, which is described with center D, and radius DA, or DC, will meet in some point M

ulterius protractae. Igitur junctis AM, et CM, erit angulus AMC acutus, quia minor (ex eadem 21. primi) illo angulo ABC, qui ponitur rectus. Quare (ex praecedente) stabilietur hypothesis anguli acuti. Quod erat tertio loco demonstrandum. Itaque constant omnia proposita. [28]

PROPOSITIO XX.

Esto triangulum ACM (fig. 19.) rectangulum in C. Tum ex puncto B dividente bifariam ipsam AM demittatur ad AC perpendicularis BD. Dico hanc perpendicularem majorem non fore (in hypothesi anguli acuti) medietate perpendicularis MC.

Demonstratur. Continuetur enim DB usque ad DH duplam ipsius DB. Foret igitur DH (si DB major sit praedicta medietate) major ipsa CM, ac propterea aequalis cuidam continuatae CMK. Jungantur AH, HK, HM, MD. Jam sic progredimur. Quoniam in triangulis HBA, DBM, aequalia ponuntur latera HB, BA, lateribus DB, BM; suntque (ex 15. primi) aequales anguli ad punctum B; erit etiam (ex quarta ejusdem primi) basis HA aequalis basi MD. Deinde, propter eandem rationem, aequales erunt in triangulis HBM, DBA, bases HM, DA. Quare in triangulis MHA, ADM, aequales erunt (ex 8. primi) anguli MHA, ADM. Rursum in triangulis AHB, MDB, aequalis manebit angulus residuus MHB residuo recto angulo ADB. Igitur rectus erit angulus MHB. At hoc

this DB produced outwardly. Therefore AM and CM being joined, the angle AMC will be acute, because less (Eu. I. 21) than the angle ABC, which is assumed right.

Therefore (from the preceding proposition) is established the hypothesis of acute angle.

Quod erat tertio loco demonstrandum.

Itaque constant omnia proposita. [28]

PROPOSITION XX.

Let there be a triangle ACM (fig. 19) right-angled at C. Then from the point B bisecting this AM let fall the perpendicular BD to AC. I say this perpendicular will not be (in the hypothesis of acute angle) greater than half the perpendicular MC.

PROOF. For let DB be produced to DH double DB. Therefore DH would be (if DB be greater than the aforesaid half) greater than CM, and therefore equal to a certain continuation CMK.

Join AH, HK, HM, MD. Now we proceed thus. Since in the triangles HBA, DBM, the sides HB, BA are assumed equal to the sides DB, BM; and (Eu. I. 15) the angles at the point B are equal;

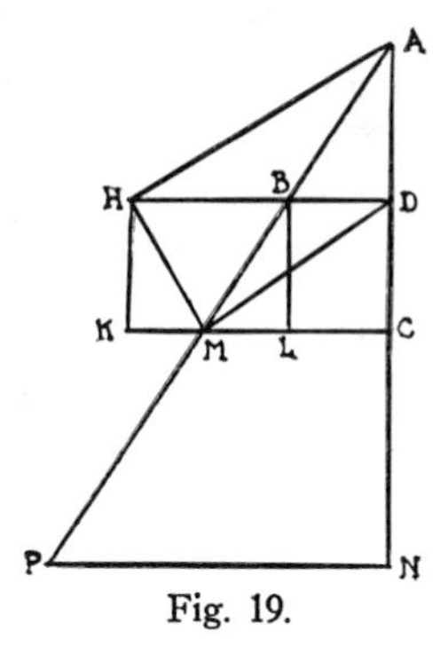

Fig. 19.

the base HA also (Eu. I. 4) will be equal to the base MD. Then, by the same reasoning, in the triangles HBM, DBA, the bases HM, DA will be equal. Wherefore in the triangles MHA, ADM, the angles MHA, ADM (Eu. I. 8) will be equal.

Again in the triangles AHB, MDB, the residual angle MHB will remain equal to the residual right angle ADB. Therefore the angle MHB will be right. But this is ab-

absurdum est, in hypothesi anguli acuti; cum recta KH jungens aequalia perpendicula KC, HD, acutos angulos efficiat cum eisdem perpendiculis. Non ergo perpendicularis BD major est (in hypothesi anguli acuti) medietate perpendicularis MC. Quod erat demonstrandum.

PROPOSITIO XXI.

Iisdem manentibus: Intelligantur in infinitum produci ipsae AM, et AC. Dico earundem distantiam majorem fore (in utraque hypothesi aut anguli recti, aut anguli acuti) qualibet assignabili finita longitudine. [29]

Demonstratur. In AM continuata sumatur AP dupla ipsius AM, demittaturque ad AC continuatam perpendicularis PN. Non erit (ex praecedente) in utravis praedicta hypothesi perpendicularis MC major medietate perpendicularis PN. Igitur PN saltem erit dupla ipsius MC, prout MC saltem est dupla alterius BD. Atque ita semper, si in continuata AM sumatur dupla ipsius AP, ex ejusque termino demittatur perpendicularis ad continuatam AC. Scilicet perpendicularis, quae ex AM semper magis continuata demittetur ad continuatam AC, multiplex erit determinatae BD supra quemlibet finitum assignabilem numerum. Igitur praedictarum rectarum distantia major erit (in utraque praedicta hypothesi) qualibet assignabili finita longitudine. Quod erat demonstrandum.

COROLLARIUM.

Quoniam vero hypothesis anguli obtusi, quae unice obesse hic posset, demonstrata jam est absolute falsa;

surd in the hypothesis of acute angle; since the straight KH joining equal perpendiculars KC, HD, makes[1] acute angles with these perpendiculars.

Therefore the perpendicular BD is not (in the hypothesis of acute angle) greater than the half of the perpendicular MC. ✡

Quod erat demonstrandum.

PROPOSITION XXI.

The same remaining: If AM and AC are understood as produced in infinitum I say their distance (in either the hypothesis of right angle, or of acute angle) will be greater than any assignable finite length. [29] ✡

Proof. In AM produced assume AP double of AM, and let fall to AC produced the perpendicular PN.

The perpendicular MC will not be (from the preceding) in either hypothesis aforesaid greater than half the perpendicular PN. Therefore PN will be at least double MC, just as MC is at least double BD.

And so always, if in AM produced is assumed double AP, and from the terminus of this a perpendicular is let fall to AC produced.

It is obvious that the perpendicular, which from AM ever more produced is let fall to AC produced, will be a multiple of the determinate BD beyond any finite assignable number.

Therefore the distance of the aforesaid straights will be (in either aforesaid hypothesis) greater than any assignable finite length.

Quod erat demonstrandum.

COROLLARY.

But since the hypothesis of obtuse angle, which alone could hinder here, is already proved absolutely false;

[1] Propp. I., VII., and XVI.

consequitur sane absolute verum esse, quod distantia unius ab altera praedictarum rectarum, si in infinitum producantur, major sit qualibet finita assignabili longitudine.

SCHOLION I.

In quo expenditur conatus Procli.

Post Theoremata a me huc usque demonstrata sine ulla dependentia ab illo Pronunciato Euclidaeo, ad cujus nempe exactissimam demonstrationem omnia conspirant; operae pretium facturum me judico, si quorundam etiam celebriorum Geometrarum labores in eandem me-[30]tam contendentium diligenter expendam. Incipio a Proclo, cujus est apud Clavium in Elementis post XXVIII. Libri primi sequens assumptum: *Si ab uno puncto duae rectae lineae angulum facientes infinite producantur, ipsarum distantia omnem finitam magnitudinem excedet.* At Proclus demonstrat quidem (ut ibi optime advertit Clavius) duas rectas (fig. 20.) ut puta AH, AD ab eodem puncto A exeuntes versus easdem partes, semper magis, in majore distantia ab eo puncto A, inter se distare, sed non etiam ita ut ea distantia crescat ultra omnem finitum designabilem limitem, prout opus foret ad ipsius intentum. Quo loco praefatus Clavius affert exemplum Conchoidis Nicomedeae, quae cum recta AH ex eodem puncto A versus easdem partes exiens, ita semper magis ab eadem recedit, ut tamen ipsarum distantia non nisi ad infinitam earundem productionem, aequalis sit cuidam finitae rectae AB perpendiculariter insistenti ipsis AH, BC, versus easdem partes in infinitum protractis. Quid

so of course follows as absolutely true, that the distance of one from the other of the aforesaid straights, if they be produced *in infinitum,* is greater than any finite assignable length.

SCHOLION I.

In which is weighed the endeavor of Proclus.

After the theorems so far demonstrated by me, independently of the Euclidean postulate, toward an exact proof of which they all conspire; in my judgment it is well if I diligently weigh the labors of certain well-known geometers in the same endeavor. [30]

I begin from Proclus, of whom Clavius in the *Elements,* after I. 28, gives the following assumption:

If from a point two straight lines making an angle are produced infinitely, their distance will exceed every finite magnitude.

But Proclus demonstrates indeed (as Clavius there well remarks) that two straights (fig. 20) as suppose AH, AD going out from the same point A toward the same parts, always diverge the more from each other, the greater the distance from the point A, but not also that this distance increases beyond every finite limit that may be designated, as was requisite for his purpose.

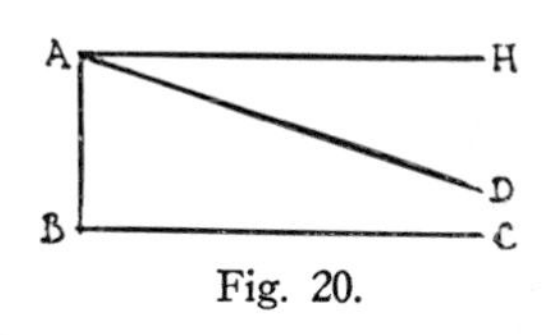

Fig. 20.

In which place the aforesaid Clavius cites the example of the Conchoid of Nicomedes, which going out from the same point A as the straight AH toward the same parts, so recedes always more from it, that nevertheless only at an infinite production is their distance equal to a certain finite sect AB standing perpendicular to AH and BC produced *in infinitum* toward the same parts. Why

ni ergo, nisi specialis ratio in contrarium cogat, dici idem possit de duabus suppositis rectis lineis AH, AD?

Neque hic accusari potest Clavius, quod Proclo opponat eam Conchoidis proprietatem, quae nempe demonstrari non potest sine adjumento plurium Theorematum, Pronunciato hic controverso innixorum. Nam dico ex hoc ipso confirmari vim redargutionis Clavianae; quia scilicet ex illo Pronunciato assumpto ut vero manifeste consequitur, duas lineas in infinitum protractas, unam rectam, et alteram inflexam, posse unam ab altera semper magis recedere intra quendam finitum determinatum limitem; unde utique oriri potest suspicio, ne simile quidpiam contingere possit in duabus lineis rectis, nisi aliter demonstretur.

Sed non idcirco; postquam ego in Cor. praecedentis [31] Propositionis manifestam jam feci absolutam veritatem praecitati assumpti; transiri statim potest ad asserendum Pronunciatum illud Euclidaeum. Nam antea demonstrari etiam oporteret, quod duae illae rectae AH, BC, quae cum incidente AB duos ad easdem partes angulos efficiant duobus rectis aequales, ut puta utrunque rectum, non etiam ipsae, ad eas partes in infinitum protractae, semper magis invicem dissiliant ultra omnem finitam assignabilem distantiam. Quatenus enim partem affirmativam praesumere quis velit; quae utique verissima est in hypothesi anguli acuti; non erit sane legitimum consequens, quod recta AD quomodolibet secans angulum HAB; unde nempe minores fiant duobus rectis duo simul ad easdem partes interni anguli DAB, CBA; quod, inquam, ea recta AD, in infinitum producta coire tandem debeat cum producta BC; etiamsi alias demon-

may not the same be said of the two assumed straight lines AH, AD, unless a special reason constrains to the contrary?

Nor here can Clavius be blamed that he opposes to Proclus this property of the Conchoid, which cannot be demonstrated except with the aid of many theorems resting upon the here controverted postulate.

For I say from this itself the force of the Clavian rebuttal is confirmed; because it is certain, this postulate being assumed, it manifestly follows, that two lines *in infinitum* protracted, one straight and the other curved, can recede one from the other ever more within a certain finite determinate limit; whence at any rate may arise a suspicion lest the same may happen for two straight lines, unless otherwise demonstrated.

But it is not therefore possible, when I now have made manifest in the corollary to the preceding [31] proposition the absolute truth of the aforesaid assumption, immediately to go over to the assertion of the Euclidean postulate.

For previously must also be demonstrated, that those two straights AH, BC, which with the transversal AB make two angles toward the same parts equal to two right angles, as for example each a right angle, do not also, protracted toward these parts *in infinitum,* always separate more from one another beyond all finite assignable distance.

For if one chooses to presume the affirmative, which is indeed entirely true in the hypothesis of acute angle; it certainly will not be a legitimate consequence, that a straight AD in any way cutting the angle HAB, hence of course making at the same time two internal angles DAB, CBA toward the same parts less than two right angles; that, I say, this straight AD, produced *in infinitum,* must at length meet with BC produced; even if it

stratum sit, quod distantia duarum AH, AD in infinitum productarum major semper evadat ultra omnem finitum designabilem limitem.

Quod autem praefatus Clavius satis esse judicaverit veritatem illius assumpti ad demonstrandum Pronunciatum hic controversum; condonari id debet praeconceptae ab ipso Clavio opinioni circa rectas lineas aequidistantes, de quibus in sequente Scholio commodius agemus.

SCHOLION II.

In quo expenditur idea Clarissimi Viri Joannis Alphonsi Borelli in suo Euclide Restituto.

Accusat doctissimus hic Auctor Euclidem, quod rectas lineas parallelas eas esse definiverit, *quae in eodem plano existentes non concurrunt ad utrasque partes, licet in infinitum producantur.* Rationem accusationis affert, quod talis [32] passio ignota sit: *tum quia,* inquit, *ignoramus, an tales lineae infinitae non concurrentes reperiri possint in natura: tum etiam quia infiniti proprietates percipere non possumus, et proinde non est evidenter cognita passio ejusmodi.*

Sed pace tanti Viri dictum sit: Numquid reprehendi potest Euclides, quod *quadratum* (ut unum inter innumera exemplum proferam) definiverit esse *figuram quadrilateram, aequilateram, rectangulam*; cum dubitari possit, an figura ejusmodi locum habeat in natura? Reprehendi, inquam, aequissime posset; si, ante omnem Problematicam demonstrativam constructionem, figuram praedictam assumpsisset tanquam datam. Hujus autem vitii immunem esse Euclidem ex eo manifeste liquet, quod nusquam praesumit quadratum a se definitum, nisi post Prop. 46. Libri primi, in qua problematice docet, ac

were at another time demonstrated, that the distance of the two AH, AD produced *in infinitum* goes out ever greater beyond all finite assignable limit.

But that the aforesaid Clavius should have judged the truth of this assumption sufficient for demonstrating the postulate here in question; that ought to be blamed to the opinion preconceived by Clavius about equidistant straight lines, which we may discuss more conveniently in a subsequent scholion.

SCHOLION II.

In which is weighed an idea of that brilliant man Giovanni Alfonso Borelli in his Euclides Restitutus.

This most learned author blames Euclid, because he defines parallel straight lines to be those, *which being in the same plane do not meet on either side, even if produced in infinitum.* ☼

He offers as ground for his accusation, that such relation [32] is unknown: *first,* he says, *because we are ignorant, whether such infinite non-concurrent lines can be found in nature: then also because we cannot perceive the properties of the infinite, and hence a relation of this sort is not clearly cognized.*

But with reverence for so great a man it may be said: Can Euclid be blamed, because (to bring forward one among innumerable examples) he defines *a square* to be *a figure quadrilateral, equilateral, rectangular*;☼ when it may be doubted, whether a figure of this sort has place in nature? He could, say I, most justly have been blamed, if, before as a problem demonstrating the construction, he had assumed the aforesaid figure as given.

But that Euclid is free from this fault follows manifestly from this, that he nowhere assumes the square defined by him, except after Prop. 46 of the First Book,

demonstrat *quadrati prout ab ipso definiti, a data recta linea descriptionem.* Simili igitur modo reprehendi nequit Euclides, quod rectas lineas parallelas eo tali modo definiverit, cum eas nusquam ad constructionem ullius Problematis assumat tanquam datas, nisi post Prop. 31. lib. primi, in qua Problematice demonstrat, quo pacto *a dato extra datam rectam lineam puncto duci debeat recta linea eidem parallela,* et quidem juxta definitionem ab eo traditam parallelarum, *ita ut nempe in infinitum protractae in neutram partem sibi invicem occurrant*: Quodque amplius est; id ipsum demonstrat sine ulla dependentia a Pronunciato hic controverso. Itaque Euclides sine ulla petitione principii demonstrat *reperiri posse in natura duas tales lineas rectas, quae* (in eodem plano consistentes) *in utramque partem in infinitum protractae nunquam concurrant*; ac propterea *cognitam nobis evidenter facit eam passionem,* per quam rectas lineas parallelas definit.

Pergamus porro, quo nos invitat diligens Euclidis accusator. Parallelas rectas lineas appellat duas quaslibet [33] rectas AC, BD, quae perpendiculariter ad easdem partes (fig. apud me 21.) insistant uni cuidam rectae AB. Nihil moror, quin definitio ejusmodi exposita sit *per passionem* (ut ipse ait) possibilem, et evidentissimam; cum (ex undecima primi) a quolibet in data recta puncto excitari possit perpendicularis.

Verum hanc ipsam et possibilitatem, et evidentiam jam demonstravi circa definitionem traditam ab Euclide. Quare unice restat, ut conferatur notum illud Pronunciatum Euclidaeum cum altero itidem Pronunciato, quod

in which in form of a problem he teaches, and demonstrates *the description from a given straight line, of the square as defined by him.*

In the same way therefore Euclid ought not to be blamed, because he defined parallel straight lines in this manner, since he nowhere assumes them as given for the construction of any problem, except after Prop. 31 of the First Book, in which as a problem he demonstrates, how *should be drawn from a given point without a given straight line a straight line parallel to this,* and indeed according to the definition of parallels given by him, *so that produced indeed into the infinite on neither side do they meet one another.* And what is more; he demonstrates this without any dependence from the postulate here controverted. Thus Euclid demonstrates without any *petitio principii* that *there can be found in nature two such straight lines, which* (lying in the same plane) *protracted on each side into the infinite never meet,* and therefore *makes clearly known to us that relation* by which he defines parallel straight lines.

Let us continue onward, whither the scrupulous accuser of Euclid invites us. Parallel straight lines he calls any two[33] straights AC, BD, which toward the same parts stand at right angles to a certain straight AB (fig. with me 21).✲ I admit that such a definition is set forth *by a state* (as he says) possible and most evident; since (Eu. I. 11) from any point in the given straight a perpendicular can be erected.

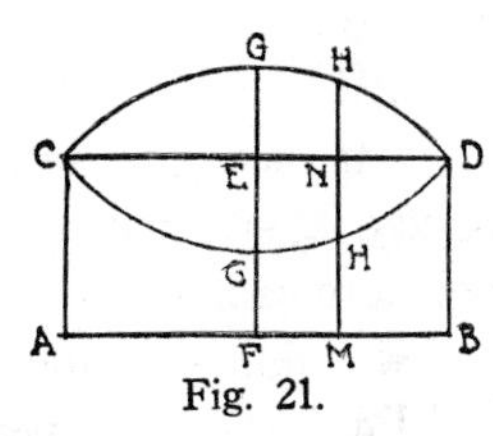

Fig. 21.

But precisely both this possibility and clearness I have just now demonstrated about the definition propounded by Euclid.

Wherefore remains only to compare that known postulate of Euclid with the other like postulate, which

usui esse debeat ad ulteriorem progressum post novam istam parallelarum definitionem. Ecce autem alterum istud Pronunciatum apud Clavium (ad quem diserte provocat ipse Borellius) in Scholio post Prop. 28. lib. primi: Si recta linea, ut puta BD super aliam rectam, ut puta BA, in transversum moveatur constituens cum ea in suo extremo B angulos semper rectos, describet alterum illius extremum D lineam quoque rectam DC, dum nempe ipsa BD pervenerit ad congruendum alteri aequali AC.

Agnosco opportunitatem Pronunciati, ut inde transitus fiat ad demonstrandum illud alterum Euclidaeum, quo nempe fulciri tandem debet reliqua omnis Geometria. Nam antea proposuerat Clavius; quod linea, cujus omnia puncta aeque distent a quadam supposita recta AB; qualis utique est (ex hypothesi praedictae descriptionis) linea DC; debet esse etiam ipsa linea recta; quia nempe ejusmodi erit, ut omnia ipsius puncta intermedia *ex aequo jaceant* (qualis est rectae lineae definitio) inter ejus extrema puncta D, et C; *ex aequo,* inquam, *jaceant*; cum omnia aeque distent ab ea supposita recta AB, nimirum quanta est longitudo ipsius BD, aut AC. Quo loco affert Clavius exemplum lineae circularis, de qua commodius infra disseremus; ubi ostendam clarissimam hac in parte disparita-[34]tem inter lineam rectam, et circularem. Nam interim dico non satis liquere, an linea descripta ab eo puncto D sit potius recta DC, quam curva quaedam DGC seu convexa, seu concava versus partes ipsius BA.

Si enim ex puncto F dividente bifariam ipsam BA intelligatur educta perpendicularis, quae occurrat rectae DC in E, et praedictis curvis in G, et G, constat sane (ex 2. hujus) rectos fore angulos hinc inde ad punctum

must be used for farther progress after the new definition of parallels.

But behold this other postulate in Clavius (to whom Borelli himself expressly refers) in the scholion after Prop. 28 of the First Book: If a straight line, as suppose BD upon another straight, as suppose BA, moves transversely making with it at its extremity B always right angles, its other extremity D describes a line also straight DC, until this BD shall have come to congruence with the other equal sect AC. I acknowledge the fitness of the postulate, that thence a transit may be made to demonstrating that other Euclidean postulate, upon which certainly at length must be supported all remaining geometry. For Clavius had previously declared; that a line, of which all points are equally distant from a certain assumed straight AB; as assuredly is (from the hypothesis of the aforesaid construction) the line DC; this line also must be straight; because certainly it will be of such sort, that all its intermediate points lie *ex aequo* (such is the definition of a straight line) between its extreme points D, and C; lie *ex aequo,* say I, since all are equally distant from this assumed straight AB, truly by as much as the length is of this BD, or AC. In this place Clavius introduces the example of the circular line, of which we shall speak more conveniently below; where I shall show the clearest disparity in this regard [34] between the straight line and circle.

But meanwhile I say it is not sufficiently evident, whether the line described by this point D is rather the straight DC than a certain curve DGC either convex or concave toward the side of this BA.

For if from the point F bisecting this BA a perpendicular is supposed erected, which meets the straight DC in E, and the aforesaid curves in G, and G, it follows surely (from P. II.) that the angles at the point E will

E; qualiscunque tandem in eo motu intelligatur descripta linea DC a puncto D; ac praeterea (ex facili intellecta superpositione) aequales hinc inde fore angulos ad puncta G, prout alterutra curva DGC descripta fuerit.

Sed rursum; assumpto in AB quolibet puncto M; si educatur perpendicularis, quae occurrat rectae DC in N, et praedictis curvis in H, et H, paulo post demonstrabo rectos fore angulos hinc inde ad punctum N, quatenus quidem recta ipsa DC genita supponatur in suo illo motu a puncto D, seu quatenus recta MN aequalis censeatur ipsi BD. Sin vero alterutra curva DHC genita putetur; ex facili itidem praescripta superpositione demonstrabitur aequales rursum hinc inde fore angulos MHD, MHC, ubivis in ea alterutra descripta curva sumptum fuerit punctum H, ex quo ad subjectam rectam lineam AB demissa intelligatur perpendicularis HM. Verum hac de re fusius, ac diligentius in altera parte hujus libri, ubi locum proprium habet.

Quorsum igitur, inquies, praecox ista anticipatio? In eum, inquam, finem; ut ne ex ista lineae eo modo genitae verissima, et a me exactissime in praecitato loco demonstranda proprietate; et quidem citra omnem defectum quomodolibet infinite parvum; praecipitanter censeremus non nisi rectam lineam esse posse. Scilicet hic inquiritur penitior rectae lineae natura, sine qua vix infantiam prae[35]tergressa Geometria subsistere ibi deberet. Non igitur hac in re vituperari potest major quaedam exactissimae veritatis inquisitio.

Neque tamen hic renuo, quin diligentissima aliqua experientia physica deprehendi possit, quod linea DC eo

be right, whatever line DC is understood at length as described in this motion by the point D; and moreover (from an easily understood superposition) the angles at the points G will be equal according as the one or the other curve DGC may be described. ✲

But again; any point M in AB being assumed; if a perpendicular is erected, which meets the straight DC in N, and the aforesaid curves in H and H, I shall prove a little later that the angles on both sides at the point N will be right, in so far indeed as this straight DC is supposed generated by the point D in that motion of its, or in as far as the straight MN is decided equal to this BD.

But if one or the other curve DHC is supposed generated; from the like aforesaid easy superposition will be demonstrated that again the angles MHD, MHC on both sides will be equal, wherever in the one or the other described curve the point H may be assumed, from which to the underlying straight line AB the perpendicular HM is understood as let fall. But of this thing more fully and more scrupulously in the Second Part of this Book, where it has its proper place.

To what end therefore, will you say, this untimely anticipation?

To this end, say I; lest from this indubitable property of the line generated in this manner, proved by me most rigorously in the aforesaid place; and indeed beyond any defect of any sort infinitely small; we may decide precipitately that the line can be only the straight.

Obviously the nature of the straight line must here be investigated more profoundly, without which geometry scarcely grown beyond infancy [35] must there remain. Therefore in this affair cannot be blamed a certain greater investigation of a most exact verity. ✲

Nor yet do I here deny, but that by some most accurate physical experimentation may be discovered, that

motu genita non nisi recta linea censenda sit. Sed quatenus ad experientiam physicam provocare hic liceat; tres statim afferam demonstrationes Physico-Geometricas ad comprobandum Pronunciatum Euclidaeum. Ubi non loquor de experientia physica tendente in infinitum, ac propterea nobis impossibili; qualis nempe requireretur ad cognoscendum, quod puncta omnia junctae rectae DC aequidistent a recta AB, quae supponitur in eodem cum ipsa DC plano consistens. Nam mihi satis erit unicus individuus casus; ut puta, si juncta recta DC, assumptoque uno aliquo ejus puncto N, perpendicularis NM demissa ad subjectam AB comperiatur esse aequalis ipsi BD, sive AC. Tunc enim anguli hinc inde ad punctum N aequales forent (ex 1. hujus) angulis sibi correspondentibus ad puncta C, et D, qui rursum (ex eadem 1. hujus) aequales inter se forent. Quare anguli hinc inde ad punctum N, atque ideo etiam reliqui duo recti erunt. Igitur unum habebimus casum pro hypothesi anguli recti; ac propterea (juxta quintam, et decimamtertiam hujus) demonstratum habebimus Pronunciatum Euclidaeum. Atque haec esse potest prima demonstratio Physico-Geometrica.

Transeo ad secundam. Esto semicirculus, cujus centrum D, et diameter AC. Si ergo (fig. 17.) in ejus circumferentia assumatur punctum aliquod B, ad quod junctae AB, CB comperiantur continere angulum rectum, satis erit hic unicus casus (prout demonstravi in 18. hujus) ad stabiliendam hypothesim anguli recti, ac propterea (ex praedicta 13. hujus) ad demonstrandum notum illud Pronunciatum. [36]

Superest tertia demonstratio Physico-Geometrica,

the line DC generated by this motion can only be adjudged a straight line.

But in so far as may be here permissible to cite physical experimentation, I forthwith bring forward three demonstrations physico-geometric to sanction the Euclidean postulate. ✡

Therewith I do not speak of physical experimentation extending into the infinite, and therefore impossible for us; such as of course would be requisite to the cognizing, that all points of the straight join DC are equidistant from the straight AB, which is supposed to be in the same plane with this DC.

For a single individual case will be sufficient for me; as suppose, if, the straight DC being joined, and any one point of it N being assumed, the perpendicular NM let fall to the underlying AB is ascertained to be equal to BD or AC. For then the angles on both sides at the point N would be equal (P. I.) to the angles corresponding to them at the points C and D, which again (from the same P. I.) would be equal *inter se*. Wherefore the angles on both sides at the point N, and therefore also the remaining two will be right.

Therefore we shall have a case for the hypothesis of right angle; and consequently (by Propp. V. and XIII.) we shall have demonstrated the Euclidean postulate. And this may be the first demonstration physico-geometric.

I pass over to the second. Let there be a semi-circle, of which the center is D, and diameter AC. If then (fig. 17) any point B is assumed in its circumference, to which AB, CB joined are ascertained to contain a right angle, this single case will be sufficient (as I have demonstrated in P. XVIII.) for establishing the hypothesis of right angle, and consequently (from the aforesaid P. XIII.) for demonstrating that famous postulate. [36]

There remains the third demonstration physico-geo-

quam puto omnium efficacissimam, ac simplicissimam, utpote quae subest communi, facillimae, paratissimaeque experientiae. Si enim in circulo, cujus centrum D, tres coaptentur (fig. 22.) rectae lineae CB, BL, LA, aequales singulae radio DC, comperiaturque juncta AC transire per centrum D, satis id erit ad demonstrandum intentum. Nam junctis DB, DL, tria habebimus triangula, quae (ex 8. et 5. primi) tum inter se invicem, tum etiam in se ipsis singula erunt aequiangula. Quoniam igitur tres simul anguli ad punctum D, nimirum ADL, LDB, BDC aequales sunt (ex 13. primi) duobus rectis; duobus etiam rectis aequales erunt tres simul anguli cujusvis illorum triangulorum, ut puta trianguli BDC. Quare (ex 15. hujus) stabilita hinc erit hypothesis anguli recti; ac propterea (ex jam nota 13. hujus) demonstratum manebit illud Pronunciatum.

Sin vero, ante omnem attentatam seu demonstrationem, seu figuralem exhibitionem, conferre inter se placeat duo illa Pronunciata, fateor sane Euclidaeum videri posse obscurius, aut etiam falsitati obnoxium. At post figuralem exhibitionem, quam Scholio IV. consequenti reservo, constabit viceversa Pronunciatum quidem Euclidaeum retinere posse dignitatem, ac nomen Pronunciati, alterum vero inter Theoremata computari tutius debere.

Sed hic explicare debeo (prout paulo ante me facturum spopondi) manifestam isto in genere disparitatem inter lineam circularem, et lineam rectam. Disparitas autem ex eo oritur; quod recta quidem linea dicitur ad se ipsam; circularis vero, ut puta (fig. 23.) MDHNM, non ad

metric, which I think the most efficacious and most simple of all, inasmuch as it rests upon an accessible, most easy, and most convenient experiment.

For if in a circle, whose center is D, are fitted (fig. 22) three straight lines CB, BL, LA, each equal to the radius DC, and it is ascertained that the join AC goes through the center D, this will be sufficient for demonstrating the assertion.

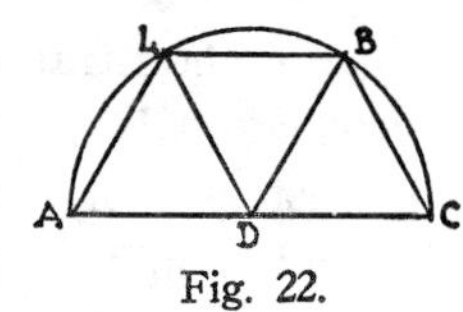

Fig. 22.

For, DB, DL being joined, we will have three triangles, which (from Eu. I. 8 and 5) not only will be equiangular to one another, but also singly for themselves. Therefore since the three angles together at the point D, indeed ADL, LDB, BDC are equal (by Eu. I. 13) to two right angles; also the three angles together of each of these triangles will be equal to two right angles, as suppose of the triangle BDC. Wherefore (from P. XV.) will be established hence the hypothesis of right angle; and consequently (from the already admitted P. XIII.) that postulate will be demonstrated.

But if, before all attempt whether at demonstration or at graphic exhibition, one wishes to compare *inter se* those two postulates, I grant indeed the Euclidean may appear more obscure or even liable to objection. But after the graphic exhibition which I reserve for Scholion IV following, it will appear *vice versa* that the Euclidean postulate indeed can retain the dignity and name of postulate, but the other ought rather to be reckoned among the theorems.

But here I must explain (as a little above I have promised I was about to do) the manifest disparity in this relation between the circular line and the straight line. Now the disparity arises from this; that a line is called straight in reference to itself; but is called circular, as suppose (fig. 23) MDHNM, not in reference to itself,

se ipsam, sed ad alterum dicitur, nimirum ad quoddam alterum in eodem cum ipsa plano existens punctum A, quod est ejusdem centrum. Consequens igitur est, prout opti-[37]me demonstratur a Clavio, quod linea FBCL in eodem cum illa plano consistens, et cujus omnia puncta aequidistent a praedicta MDHNM, sit et ipsa circularis, nimirum omnibus suis punctis aequidistans a communi centro A. Quod enim BD, quae sit continuatio in rectum ipsius AB, sit mensura distantiae illius puncti B ab ea circulari MDHNM, ex eo constat; quia (ex 7. tertii, quae est independens a Pronunciato hic controverso) minima omnium ipsa est, quae ab eo puncto in eam circumferentiam cadere possint. Idem valet de reliquis CH, LN, FM. Quoniam igitur et totae AM, AD, AH aequales sunt, utpote radii ex centro A ad suppositam lineam circularem MDHNM; atque item aequales sunt abscissae FM, BD, CH, LN, quae nempe mensura sunt aequalis distantiae omnium punctorum illius lineae FBCLF ab ea supposita linea circulari MDHNM; consequens plane est, ut aequales pariter sint residuae AF, AB, AC, AL, ac propterea ipsa etiam linea FBCLF sub eodem centro A circularis sit.

Numquid autem uniformiter, ad demonstrandum, quod linea DC (fig. 21.) eo tali motu genita a puncto D sit linea recta, satis erit aequidistantia omnium ipsius punctorum a subjecta recta AB? Nullo modo. Nam linea recta dicitur absolute ad se ipsam, sive in se ipsa, nimirum ita *ex aequo jacens inter sua puncta,* ac praesertim extrema, ut manentibus istis immotis nequeat ipsa revolvi ad occupandum novum locum. Nisi haec passio aliquo pacto demonstretur de ea DC, nunquam constabit eam

but to something else, forsooth to a certain other point A existing in the same plane with it, which is its center.

The consequence therefore is, as is most excellently [37] demonstrated by Clavius, that the line FBCL existing in the same plane with it, and whose points are all equidistant from the aforesaid MDHNM, is also itself circular, truly equidistant in all its points from the common center A. That in fact BD, which is the continuation in a straight of AB, is the measure of the distance of that point B from this circle MDHNM follows from this; because (from Eu. III. 7, which is independent of the postulate here in controversy) this is the smallest of all, which can fall from this point upon this circumference. The same holds of the remaining CH, LN, FM.

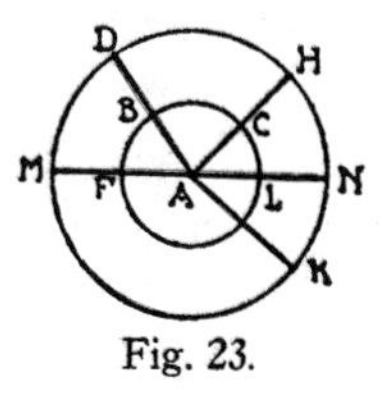

Fig. 23.

Since therefore also the wholes AM, AD, AH, are equal as radii from the center A to the line assumed circular MDHNM; and also the sections FM, BD, CH, LN are equal, which obviously are the measure of the equal distance of all points of that line FBCLF from this line presumed circular MDHNM; the consequence plainly is, that equal likewise are the remainders AF, AB, AC, AL, and therefore also this line FBCLF is a circle with the same center A.

But now likewise, for demonstrating that the line DC (fig. 21) generated through such a motion by the point D is a straight line will the equidistance of all its points from the underlying straight AB be sufficient? In no way.

For a line is called straight absolutely in reference to itself, or in itself, doubtless as *lying ex aequo between its points,* and especially end points, so that these remaining unmoved it cannot be revolved into occupying a new place.✲ Unless this state in some way be demonstrated of this

esse lineam rectam, qualiscunque tandem supponatur, aut demonstretur omnium ipsius punctorum relatio ad subjectam in eodem plano rectam AB; praesertim vero, ne uniformiter dicamus nullam aliam in eo plano fore lineam rectam, quae omnibus suis punctis non aequidistet ab ea supposita recta linea AB. [38]

Neque tamen dictum hoc meum ita accipi volo, quasi putem demonstrari non posse, quod linea sic genita ipsa sit linea recta, nisi post demonstratam veritatem controversi Pronunciati; cum magis ego ipse prope finem hujus Libri demonstraturus id sim, ad confirmandum ipsum tale Pronunciatum.

SCHOLION III.

In quo expenditur conatus Nassaradini Arabis, et simul idea super eodem negotio Clariss. Viri Joannis Vallisii.

Conatum istum Nassaradini Arabis latino idiomate typis vulgavit praelaudatus Vir Joannes Vallisius, cum animadversionibus opportuno loco adjectis. Duo autem in rem suam postulat sibi concedi Nassaradinus.

Primum est; ut duae quaelibet rectae lineae in eodem plano positae, in quas aliae quotlibet rectae lineae ita incidant, ut uni quidem earum perpendiculares semper sint, alteram vero ad angulos inaequales semper secent, nimirum versus unam partium sub angulo semper acuto, et versus alteram sub angulo semper obtuso; ut, inquam, priore loco dictae lineae censeantur semper magis (quandiu se mutuo non secent) ad se invicem accedere versus partes illorum angulorum acutorum; et vicissim semper magis a se invicem recedere versus partes angulorum obtusorum.

At ego quidem, si nihil aliud moratur Nassaradinum,

DC it will never be certain that this is a straight line, whatever relation finally is supposed or demonstrated of all its points to the underlying straight AB in the same plane; but especially we must not say analogically that no other line in this plane will be straight which in all its points is not equidistant from this line AB supposed straight. [38]

Nor finally do I wish this dictum of mine so taken, as if I think it cannot be demonstrated, that the line thus generated is itself a straight line, except after truth demonstrated of the controverted postulate; since rather I myself will demonstrate it toward the end of this Book, for confirming such postulate itself.

SCHOLION III.

In which is weighed the endeavor of the Arab Nasiraddin, and likewise the idea of the illustrious John Wallis upon the same affair.

This endeavor of the Arab Nasiraddin the above eulogized John Wallis has published in the Latin language with remarks added in opportune place.

However Nasiraddin requires two things to be conceded to him in this affair.

The first is; that any two straight lines lying in the same plane, upon which ever so many other straight lines so strike, that they are always perpendicular to one indeed of these, but always cut the other at unequal angles, truly toward one part always under an acute angle, and toward the other always under an obtuse angle; that, I say, the above-mentioned lines be supposed always more (as long as they do not mutually cut) to approach each other toward the side of those acute angles; and on the other hand always more to recede from one another toward the parts of the obtuse angles.

But I indeed, if nothing else impedes Nasiraddin, wil-

libens permitto, quod postulat; cum istud ipsum, quod ab eo indemonstratum relinquitur, intelligi possit exactissime a me demonstratum in Cor. II. post 3. hujus.

Alterum Nassaradini Postulatum est reciprocum primi; ut nempe acutus semper sit angulus versus eas partes, [39] ad quas jam dictae perpendiculares supponantur fieri semper breviores; obtusus autem versus alias partes, ad quas eaedem perpendiculares supponantur evadere semper longiores.

Verum hic latet aequivocatio. Cur enim (dum ab una aliqua statuta tanquam prima perpendiculari procedatur ad alias) consequentium perpendicularium anguli, ad eandem partem acuti, non fiant semper majores, quo usque incidatur in angulum rectum, nimirum in talem perpendicularem, quae ipsa sit utriusque praedictarum rectarum commune perpendiculum? Et istud quidem si accidat, evanescit latebrosa ista Nassaradini praeparatio, postquam ingeniose quidem, sed magno cum labore Euclidaeum Pronunciatum demonstrat.

Quod si Nassaradinus jure quodam suo praesumere velit tanquam per se notam consistentiam illam ad eandem partem angulorum acutorum: Cur non etiam (dicam cum Vallisio) concipi potest tanquam per se clarum: *Duas rectas in eodem plano convergentes* (in quas nempe alia recta incidens duos ad easdem partes angulos efficiat minores duobus rectis, ut puta unum rectum, et alterum quomodolibet acutum) *tandem occursuras, si producantur*? Neque enim opponi potest, quod major ista ad unas partes convergentia subsistere semper possit intra quendam determinatum limitem, adeo ut nempe tanta quaedam distantia inter eas lineas ad eam partem semper intersit, etiamsi caeteroquin una ad alteram semper propius acce-

lingly permit what he postulates; since just that, which with him remains undemonstrated, can be recognized as most rigorously demonstrated by me in Cor. II. to P. III.

The other postulate of Nasiraddin is the reciprocal of the first; that indeed the angle may always be acute toward those parts [39] where the just mentioned perpendiculars are supposed to become shorter; but obtuse toward the other parts where these perpendiculars are supposed to go out always longer. But here lurks an ambiguity.

For why (while from any one perpendicular prescribed as the first we proceed to the others) may not the angles of the consequent perpendiculars, on the same side acute, not become ever greater, even to where one strikes upon a right angle, consequently upon such a perpendicular as is itself the common perpendicular to each of the aforesaid straights? And if indeed that happens, evanishes this subtle preparation of Nasiraddin, by means of which ingeniously indeed, but with great labor he demonstrates the Euclidean postulate.

And yet if Nasiraddin with a certain justice may determine to presume as if known *per se* that persistence of acute angles on the same side: why cannot also (I speak with Wallis) be assumed as if clear *per se*: *Two straights in the same plane converging* (upon which of course another straight striking makes toward the same parts two angles less than two right angles, as suppose one right, and the other in whatever way acute) *finally meet, if produced*?

Nor in fact can it be objected, that this greater convergence toward one side can always subsist within a certain determinate limit, so that indeed a certain so much of distance always intervenes between these lines on this side, even if still one approaches always more nearly to the other.

dat. Non, inquam, opponi id potest; quoniam ex hoc ipso demonstrabo, post XXV. hujus, omnium talium rectarum ad finitam distantiam occursum, juxta Pronunciatum Euclidaeum.

Jam transeo ad praelaudatum Joannem Vallisium, qui nempe, ut morem gereret tot Magnis Viris, Veteribus pariter, ac Recentioribus, et rursum ex onere Cathe-[40]drae suae Oxoniensi imposito, hoc idem pensum aggredi voluit demonstrandi saepe dictum Pronunciatum. Unice autem assumit tanquam certum, quod sequitur: nimirum *Datae cuicunque figurae similem aliam cujuscunque magnitudinis possibilem esse.* Et id quidem praesumi posse de qualibet figura (etiam si in rem suam unice assumat triangularem rectilineam) bene argumentatur ex circulo, quem scilicet sub quantolibet radio describi posse omnes agnoscunt. Deinde acutus Vir cautissime observat praesumptioni huic suae non obstare, quod praeter correspondentium angulorum aequalitatem requiratur etiam correspondentium omnium laterum proportionalitas, ut habeatur una figura rectilinea, v. g. triangularis, alteri rectilineae triangulari similis; cum tamen Proportionalium, ac subinde similium Figurarum definitio ex Quinto, ac Sexto Euclidis Libro desumendae sint: *Poterat enim Euclides* (inquit ipse) *utramque Libro Primo praemisisse.* Porro autem, hoc stante (quod tamen negari a quopiam posset, nisi demonstretur) intentum suum pulchro sane, atque ingenioso molimine exequitur laudatus Vir.

Sed nolo oneri a me suscepto in quoquam deesse. Itaque assumo duo triangula, unum ABC, et alterum DEF (fig. 24.) invicem aequiangula: Non dico plane

That cannot, I say, be objected; since from this itself I shall demonstrate, after P. XXV., the meeting at a finite distance of all such straights, in accordance with the Euclidean postulate.

Now I go over to the aforesaid John Wallis, who, as made a custom with so many great men, ancient as well as recent, and on the other hand from the obligation imposed on his Oxford professional chair, [40] determined to undertake this same duty of demonstrating the oft mentioned postulate.

Now solely he assumes as if certain, what follows: namely that *to any given figure another similar of any magnitude is possible.*

And that this indeed may be presumed of any figure (although in his affair he assumes solely a rectilineal triangle) is well argued from the circle, which of course all admit can be described with any-sized radius.

Further the acute man observes most cautiously it does not thwart this his presumption, that besides the equality of corresponding angles also the proportionality of all corresponding sides is required, in order that a rectilineal figure, for example a triangle, may be similar to another rectilineal triangle; though still the definition of proportion, and forthwith of similar figures are to be taken from the Fifth, and the Sixth Books of Euclid: *For* (says he himself) *Euclid could have put each in front of the First Book.*

Hereafter, this standing (which nevertheless can be denied by any one, unless it is demonstrated) the famous man carries out his intent with really beautiful and ingenious effort.

But I am unwilling to fail in anything to the charge undertaken by me.

Therefore I assume two triangles, one ABC, and the other DEF (fig. 24) mutually equiangular. I do not

similia; quia non indigeo proportionalitate laterum circa angulos aequales, immo neque ulla ipsorum laterum determinata mensura. Solum igitur nolo triangula invicem aequilatera, quia tunc sufficeret sola octava primi, sine ulla praesumptione. Itaque anguli ad puncta A, B, C, aequales sint angulis ad puncta D, E, F; sitque latus DE minus latere AB; assumaturque in AB portio AG aequalis ipsi DE, atque item in AC portio AH aequalis ipsi DF. Debere autem DF minorem esse ipsa AC infra declarabo. Tum (juncta GH) constat (ex 4. primi) aequales fore [41] angulos ad puncta E, et F, ipsis AGH, AHG. Quapropter; cum modo dicti anguli una cum aliis BGH, CHG, aequales sint (ex 13. primi) quatuor rectis; quatuor itidem rectis aequales erunt anguli ad puncta B, et C, una cum eisdem angulis BGH, CHG. Igitur quatuor simul anguli quadrilateri BGHC aequales erunt quatuor rectis; ac propterea (ex 16. hujus) stabilietur hypothesis anguli recti; et simul (ex 13. hujus) Pronunciatum Euclidaeum.

Porro supposui latus DF, sive AH sumptum ipsi aequale, minus fore latere AC. Si enim aequale foret, et sic punctum H caderet in punctum C; tunc angulus BCA aequalis foret (ex hypothesi) angulo EFD, sive GCA (qui tunc fieret) totum parti; quod est absurdum. Sin vero majus foret, et sic juncta GH secaret in aliquo puncto ipsam BC; jam angulus ACB externus aequalis foret ex hypothesi (contra 16. primi) angulo interno, et

say wholly similar; because I do not need the proportionality of the sides about the equal angles, nay nor any determinate measure of the sides themselves. Merely therefore I do not wish triangles mutually equilateral, since then Eu. I. 8 would alone suffice, without any assumption.

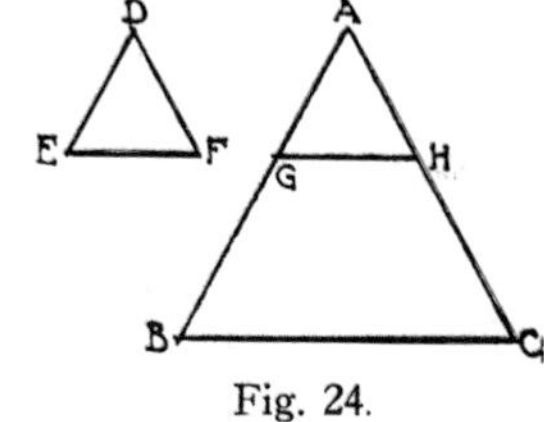

Fig. 24.

So let the angles at the points A, B, C, be equal to the angles at the points D, E, F; and let the side DE be less than the side AB; and in AB is assumed the portion AG equal to this DE, and likewise in AC the portion AH equal to this DF. But that DF must be less than AC I will make clear below. Then (GH joined) follows (from Eu. I. 4) the angles at the points E, and F will be equal [41] to AGH, AHG. However since the just mentioned angles, together with the others BGH, CHG, are equal (Eu. I. 13) to four right angles; likewise will be equal to four right angles the angles at the points B, and C, together with these same angles BGH, CHG. Therefore the four angles of the quadrilateral BGHC will be together equal to four right angles; and consequently (from P. XVI.) is established the hypothesis of right angle; and at the same time (from P. XIII.) the Euclidean postulate.

Moreover I have supposed the side DF, or AH assumed equal to it, to be less than the side AC. For if it were equal, and so the point H should fall upon the point C, then the angle BCA would be equal (by hypothesis) to the angle EFD, or GCA (which then it would become) a part to the whole; which is absurd.

But if it were greater, and so the join GH should cut BC itself in some point, now the external angle ACB would be from the hypothesis equal (against Eu. I. 16) to

opposito (qui tunc fieret) AHG, sive GHA. Itaque bene supposui latus DF unius trianguli minus fore latere AC alterius trianguli, juxta hypothesim jam stabilitam.

Quare ex duobus quibusvis invicem aequiangulis triangulis, sed non etiam invicem aequilateris, stabilitur Pronunciatum Euclidaeum. Quod intendebatur.

SCHOLION IV.

In quo exponitur figuralis quaedam exhibitio, ad quam fortasse respexit Euclides, ut suum illud Pronunciatum tanquam per se notum stabiliret.

Praemitto primo: sub quolibet angulo acuto BAX (recole ex hac Tab. Fig. 12.) educi posse ex aliquo [42] puncto X ipsius AX quandam XB, quae sub quovis designato etiamsi obtuso angulo R, qui nimirum cum eo acuto BAX deficiat a duobus rectis; quandam, inquam, educi posse XB, quae ad finitam distantiam occurrat ipsi AB in quodam puncto B. Nam id ipsum jam demonstravi in Scholio post XIII. hujus.

Praemitto secundo: eas AB, AX (fig. 25.) intelligi posse in infinitum protractas usque in quaedam puncta Y, et Z; atque item praedictam XB (in infinitum et ipsam protractam usque in quoddam punctum Y) intelligi posse

the internal and opposite angle (which then would become) AHG, or GHA. ☼

Therefore I have rightly supposed the side DF of one triangle to be less than the side AC of the other triangle, in accordance with the hypothesis now established.

Wherefore from any two triangles mutually equiangular, but not also mutually equilateral, the Euclidean postulate is established. Quod intendebatur.

SCHOLION IV. ☼

In which is expounded on a figure a certain consideration, of which Euclid probably thought, in order to establish that postulate of his as per se evident.

I premise first: within any acute angle BAX (fig. 12) can be drawn from any [42] point X of AX a certain straight XB, under any designated (even obtuse) angle R (provided only that R with the acute BAX falls short of two right angles); I say, a certain XB can be drawn, which at a finite remove meets AB in a certain point B.

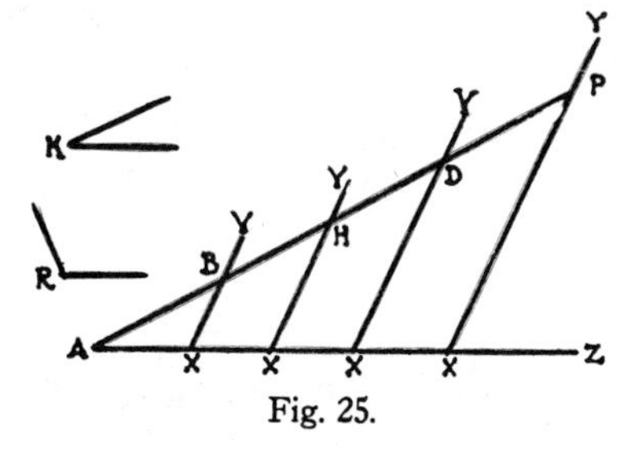

Fig. 25.

For just that I have already demonstrated in a scholion after P. XIII. ☼

I premise secondly: these AB, AX (fig. 25) can be understood as produced *in infinitum* even to certain points Y, and Z;☼ and likewise the aforesaid XB (produced *in infinitum* even to a point Y) can be understood to be

ita moveri super ea AZ versus partes puncti Z, ut angulus ad punctum X versus partes puncti A aequalis semper sit dato cuivis obtuso angulo R.

Praemitto tertio: nulli jam dubitationi obnoxium fore illud Pronunciatum Euclidaeum, si antedicta XY in eo quantocunque motu super recta AZ secet semper illam AY in quibusdam punctis B, H, D, P, atque ita consequenter in aliis punctis remotioribus ab eo puncto A. Ratio evidens est; quia sic duae quaelibet in eodem plano existentes rectae AB, XH, in quas recta quaelibet incidens AX duos ad easdem partes angulos BAX, HXA, duobus rectis minores efficiat, convenire tandem ad eas partes deberent in uno eodemque puncto H.

Praemitto quarto: nulli item dubitationi locum fore super veritate praecedentis hypothetici assumpti; si posteriores illi externi anguli YHD, YDP, et sic alii quilibet consequentes, aut aequales semper sint priori externo angulo YBD, aut saltem non ita minores semper sint, quin eorum unusquisque major semper sit parvulo quopiam designato acuto angulo K: Hoc enim stante manifestum fiet, quod ea XY, in suo illo quantocunque motu versus partes puncti Z, nunquam cessabit secare praedictam AY; quod utique (ex praecedente notato) satis est ad sta-[43] biliendum Pronunciatum controversum.

Unice igitur superest, ut quidam Adversarius dicat angulos illos externos in majore, ac majore distantia ab illo puncto A fieri semper minores sine ullo determinato limite. Inde autem fiet, ut illa XY in suo illo motu super recta AZ occurrere tandem debeat ipsi AY in quodam puncto P sine ullo angulo cum segmento PY, adeo ut nempe segmentum ejusmodi commune sit duarum recta-

so moved above this AZ toward the parts of the point Z, that the angle at the point X toward the parts of the point A is always equal to the certain given obtuse angle R.

I premise thirdly: that Euclidean postulate would be liable now to no doubt, if the aforesaid XY in this however great motion above the straight AZ cuts always that AY in certain points B, H, D, P, and so successively in other points more remote from this point A.

The reason is evident; since thus any two straights AB, XH lying in the same plane, upon which any straight AX cutting makes two angles toward the same parts BAX, HXA, less than two right angles, must at length meet toward those parts in one and the same point H.

I premise fourthly: likewise will be no doubt about the truth of the preceding hypothetical assumption, if the later external angles YHD, YDP and so any other succeeding ones, either always are equal to the preceding external angle YBD, or at least always will be not so much less but that any one of them always will be greater than any little designated acute angle K. For, this holding, it is manifest that this XY in that however great motion of its toward the parts of the point Z, never will cease to cut the aforesaid AY; which assuredly (from the preceding remark) is sufficient for establishing [43] the controverted postulate.

Solely therefore remains, that some adversary may say those external angles at greater and greater distance from the point A may become always less without any determinate limit.

But thence would follow, that XY in its motion above the straight AZ would at length meet AY in a certain point P without any angle with the segment PY, so that indeed a segment of the two straights APY, and XPY

rum APY, et XPY. At hoc evidenter repugnat naturae lineae rectae.

Sin vero cuiquam minus opportunus videatur angulus obtusus ad illud punctum X versus partes puncti A, nullo negotio supponi poterit rectus; adeo ut nempe (in motu praedictae XY ad angulos semper rectos super recta AZ) manifestius appareat singula illius XY puncta aequabiliter semper moveri relate ad subjectam AZ; ac propterea nequire jam dictam XY transire de secante in non secantem alterius indefinitae AY, nisi eam aut aliquando in aliquo puncto praecise contingat, aut ipsi occurrat in aliquo puncto P, ubi cum eadem AY commune obtineat segmentum PY; quorum utrunque adversari naturae lineae rectae ostendam ad XXXIII. hujus. Igitur juxta veram ideam lineae rectae, debebit illa XY, in quantacunque distantia puncti X a puncto A, occurrere semper in aliquo puncto ipsi AY. Atque id quidem (quantumlibet parvus supponatur acutus angulus ad punctum A) satis esse ad demonstrandum, contra hypothesim anguli acuti, Pronunciatum Euclidaeum, constabit ex XXVII. hujus.

PROPOSITIO XXII.

Si duae rectae AB, CD in eodem plano existentes perpendiculariter insistant cuidam rectae BD; ipsa autem AC jungens ea perpendicula internos (in hypothesi anguli acuti) acu-[44]*tos angulos cum eisdem efficiat: Dico* (*fig.* 26.) *rectas terminatas AC, BD commune aliquod habere perpendiculum, et quidem intra limites designatis punctis A, et C praefinitos.*

would be in this way common. But this is evidently repugnant to the nature of the straight line. ✡

But if to any one may seem less opportune the obtuse angle at the point X toward the parts of the point A, it may easily be supposed right; so that indeed (in the motion of the aforesaid XY at angles always right above the straight AZ) more manifestly may appear that the single points of that XY are always moved uniformly relatively to the basal AZ; and therefore the aforesaid XY cannot go over from a secant into a non-secant of the other indefinite AY, unless either once in some point it precisely touches it, or meets it in some point P, where it has with this AY a common segment PY; each of which I shall show contrary to the nature of the straight line in P. XXXIII.

Therefore in accordance with the true idea of the straight line, must that XY, however great the distance of the point X from the point A, always meet in some point this AY. And that this indeed (however small is supposed the acute angle at the point A) is sufficient for demonstrating, against the hypothesis of acute angle, the Euclidean postulate, will follow from P. XXVII.

PROPOSITION XXII.

If two straights AB, CD existing in the same plane stand perpendicular to a certain straight BD; but AC joining these perpendiculars makes with them internal acute angles (in hypothesis of acute angle): [44] *I say (fig. 26) the terminated straights AC, BD have a common perpendicular, and indeed within the limits fixed by the designated points A and C.*

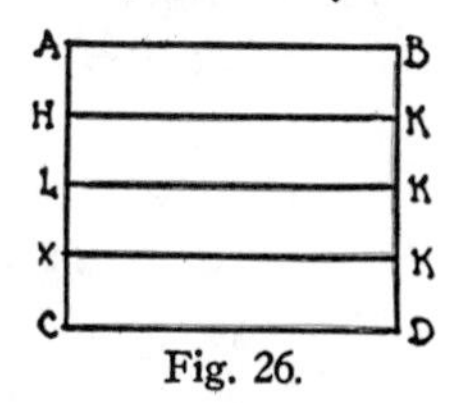

Fig. 26.

Demonstratur. Si enim aequales sint ipsae AB, CD; constat (ex 2. hujus) rectam LK, a qua bifariam dividantur illae duae AC, et BD, commune fore eisdem perpendiculum. Sin vero alterutra sit major, ut puta AB: demittatur ad BD (juxta 12. primi) ex quovis puncto L ipsius AC perpendicularis LK, occurrens alteri BD in K. Occurret autem in aliquo puncto K, consistente inter puncta B, et D; ne (contra 17. primi) perpendicularis LK secet alterutram AB, aut CD, perpendiculares eidem BD. Si ergo anguli ad punctum L recti non sunt, unus eorum acutus erit, et alter obtusus. Sit obtusus versus punctum C. Jam vero intelligatur LK ita procedere versus AB, ut semper ad rectos angulos insistat ipsi BD, atque item opportune aucta, aut imminuta, in aliquo sui puncto secet rectam AC. Constat angulos ad puncta intersectiva ipsius AC non posse omnes esse obtusos versus partes puncti C, ne tandem in ipso puncto A, dum recta LK congruet cum recta AB, angulus ad punctum A versus partes puncti C sit obtusus, cum ad eas partes positus sit acutus. Quoniam ergo angulus ad punctum L ipsius LK positus est obtusus versus partes puncti C, non transibit in eo motu recta LK ad faciendum in aliquo sui puncto cum recta AC angulum acutum versus partes praedicti puncti C, nisi prius transeat ad constituendum in aliquo sui puncto cum eadem AC angulum rectum versus partes ejusdem puncti C. Erit igitur inter puncta A, et L unum aliquod punctum intermedium H, in quo HK perpendicularis ipsi BD sit etiam perpendicularis alteri AC.

Simili modo ostendetur adesse aliquam XK inter ipsas LK, CD, quae sit perpendicularis et rectae BD, et [45]

Proof. For if AB, CD are equal, it follows (from P. II.) that the straight LK, by which these two AC and BD are bisected, will be to them a common perpendicular. But if either be the greater, as suppose AB; let fall to BD (according to Eu. I. 12) from any point L of AC the perpendicular LK, meeting the other BD in K.

But it will meet it in some point K existing between the points B and D; otherwise (contrary to Eu. I. 17) the perpendicular LK would cut either AB, or CD, perpendicular to the same BD. If then the angles at the point L are not right, one of them will be acute and the other obtuse.

Let the obtuse be toward the point C. But now LK is understood so to proceed toward AB, that it always stands at right angles to BD, and likewise opportunely increased, or diminished, in some point of it cuts the straight AC. It follows that the angles at the intersection points with AC cannot all be obtuse toward the parts of the point C, lest at length in that point A, where the straight LK is congruent with the straight AB, the angle at the point A toward the parts of the point C should be obtuse, when toward these parts it is by hypothesis acute.

Since therefore the angle at the point L of this LK is by hypothesis obtuse toward the parts of the point C, the straight LK will not change over in this motion so as to make in some point of it with the straight AC an angle acute toward the parts of the aforesaid point C, unless previously it changes over so as to make in some point of it with this AC an angle right toward the parts of this same point C.

Therefore between the points A, and L will be some one intermediate point H, in which HK perpendicular to this BD is also perpendicular to the other AC. ✡

In a similar manner is shown to be present a certain XK between LK, CD, which is perpendicular both to the

rectae AC, dum scilicet angulus obtusus ad punctum L ponatur consistere versus partes puncti A.

Constat igitur rectas AC, BD commune aliquod habituras esse perpendiculum, et quidem intra limites designatis punctis A, et C praefinitos, quoties junctae AB, CD in eodem plano existant, sintque perpendiculares ipsi BD. Quod erat etc.

PROPOSITIO XXIII.

Si duae quaelibet rectae AX, BX (fig. 27.) in eodem plano existant; vel unum aliquod (etiam in hypothesi anguli acuti) commune obtinent perpendiculum; vel in alterutram eandem partem protractae, nisi aliquando ad finitam distantiam una in alteram incidat, semper magis ad se invicem accedunt.

Demonstratur. Ex quolibet puncto A ipsius AX demittatur ad rectam BX perpendicularis AB. Si ipsa BA efficiat cum AX angulum rectum, habemus intentum communis perpendiculi. Caeterum vero ea recta efficiat ad alterutram partem, ut puta versus partes puncti X, angulum acutum. Itaque in praedicta recta AX designentur inter puncta A, et X quaelibet puncta D, H, L, ex quibus demittantur ad rectam BX perpendiculares DK, HK, LK. Si unus aliquis angulus ad puncta D, H, L acutus sit versus partes puncti A, constat (ex praecedente) unum aliquod adfuturum commune perpendiculum ipsarum AX, BX. Sin vero omnis hujusmodi angulus sit major acuto; vel unus aliquis erit rectus, et sic rursum

straight BD, and [45] to the straight AC, if namely an angle at the point L is assumed to be obtuse toward the parts of the point A.

It follows therefore that the straights AC, BD, will have a common perpendicular, and indeed within the limits fixed by the designated points A, and C, when the joins AB, CD exist in the same plane and are perpendicular to BD.

Quod erat etc.

PROPOSITION XXIII. ✡

If any two straights AX, BX (fig. 27) are in the same plane; either they have (even in the hypothesis of acute angle) a common perpendicular; or prolonged toward the same part unless somewhere at a finite distance one meets the other—they mutually approach ever more toward each other.

PROOF. From any point A of AX let fall to the straight BX the perpendicular AB. If BA makes with AX a right angle, we have the asserted case of a common perpendicular. But otherwise this straight makes toward one or the other part, as suppose toward the parts of the point X, an acute angle. Accordingly in the aforesaid straight AX between the points A and X any points D, H, L are designated, from which are let fall to the straight BX the perpendiculars DK, HK, LK.

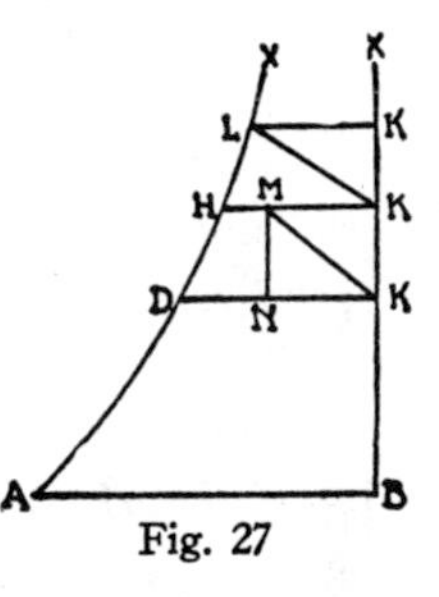

Fig. 27

If any one angle at the points D, H, L be acute toward the parts of the point A, it follows (from the preceding) that AX, BX will have a common perpendicular.

But if every angle of this sort be greater than acute; either some one will be right, and thus again we shall

habemus intentum communis perpendiculi, cum omnes anguli ad puncta K supponantur recti; vel omnes illi anguli ponuntur obtusi versus partes puncti A, ac propterea omnes itidem acuti versus partes puncti X, et sic rursum argumentor. Quoniam in quadrilatero KDHK recti sunt [46] anguli ad puncta K, ponitur autem acutus angulus ad punctum D, erit (ex Cor. II. post 3. hujus) latus DK majus latere HK. Simili modo ostendetur latus HK majus esse latere LK; atque ita semper, conferendo inter se perpendiculares ex quolibet puncto altiore ipsius AX demissas ad alteram BX. Quapropter ipsae AX, BX semper magis versus partes puncti X ad se invicem accedent: Quae est altera pars propositi disjuncti.

Ex quibus omnibus constat duas quaslibet rectas AX, BX, quae in eodem plano existant, vel unum aliquod (etiam in hypothesi anguli acuti) commune habere perpendiculum, vel in alterutram eandem partem protractas, nisi aliquando ad finitam distantiam una in alteram incidat, semper magis ad se invicem accedere. Quod erat etc.

COROLLARIUM I.

Hinc anguli versus basim AB erunt semper obtusi ad illud punctum ipsius AX, ex quo demittitur perpendicularis ad rectam BX: erunt, inquam, semper obtusi, quoties duae illae AX, et BX semper magis ad se invicem accedant versus partes punctorum X; quod quidem sano modo intelligi debet, nimirum de perpendicularibus demissis ante praedictum occursum, si forte ad finitam distantiam una in alteram incidere debeat.

have the asserted case of a common perpendicular, since all angles at the points K are supposed right; or all those angles toward the parts of the point A are obtuse, and therefore all therewith acute toward the parts of the point X, and so again I argue: Since in the quadrilateral KDHK the angles at the points K are right, [46] but the angle at the point D is acute, the side DK will be (from Cor. II. to P. III.) greater than the side HK.

In a similar way the side HK is shown to be greater than the side LK; and so always, comparing to each other perpendiculars from any ever higher points of AX let fall upon the other BX.

Wherefore AX, BX mutually approach each other ever more toward the parts of the point X: which is the second part of the disjunct proposition.

From all which follows that any two straights AX, BX, which are in the same plane, either have (even in the hypothesis of acute angle) a common perpendicular, or produced toward either the same part, unless somewhere at a finite distance one meets the other, mutually approach each other ever more.

Quod erat etc.

COROLLARY I.

Hence the angles toward the base AB will be always obtuse at each point of AX, from which is let fall a perpendicular to the straight BX: will be, I say, always obtuse, as those two AX, and BX mutually approach each other ever more toward the parts of the points X; which of course should be understood in a sane way, of perpendiculars let fall before the mentioned meeting, if perchance one is to strike upon the other at a finite distance.

SCHOLION.

Video tamen inquiri hic posse, qua ratione ostendendum sit commune illud perpendiculum; quoties recta quaepiam PFHD (fig. 28.) occurrens duabus AX, BX in punctis F, et H, duos ad easdem partes efficiat internos angulos AHF, BFH, non eos quidem rectos, sed tamen [47] aequales simul duobus rectis. Ecce autem commune illud perpendiculum geometrice demonstratum. Divisa FH bifariam in M demittantur ad AX, et BX perpendiculares MK, ML. Angulus MFL aequalis erit (ex 13. primi) angulo MHK, qui nempe supponitur duos rectos efficere cum angulo BFH. Praeterea recti sunt anguli ad puncta K, et L; ac rursum aequales sunt ipsae MF, MH. Igitur (ex 26. primi) aequales itidem erunt anguli FML, HMK. Quare angulus HMK duos efficiet rectos angulos cum angulo HML, prout cum eodem duos efficit rectos angulos (ex 13. primi) angulus FML. Igitur (ex 14. primi) una erit recta linea continuata ipsa KML, commune idcirco perpendiculum praedictis rectis AX, BX. Quod erat etc.

COROLLARIUM II.

Sed rursum docere hinc possum, quod illae duae AX, BX, in quas incidens recta PFHD, aut duos efficiat cum ipsis AX, BX internos ad easdem partes angulos aequales

SCHOLION.

I see indeed it may here be asked in what way that common perpendicular can be shown, when any straight PFHD (fig. 28) meeting two AX, BX in points F, and H, makes toward the same parts two internal angles AHF, BFH, not themselves indeed right, but nevertheless [47] together equal to two rights. But behold that common perpendicular geometrically demonstrated.

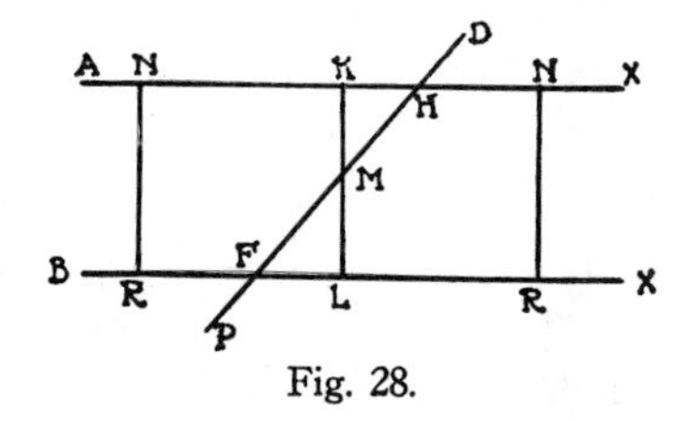

Fig. 28.

FH being bisected in M, perpendiculars MK, ML are let fall to AX and BX. The angle MFL will be equal (Eu. I. 13) to the angle MHK, which indeed is assumed to make up two right angles with the angle BFH. Moreover the angles at the points K, and L are right; and again MF, MH are equal. Therefore (Eu. I. 26) so are the angles FML, HMK equal. Wherefore the angle HMK makes two right angles with the angle HML, since with this the angle FML (Eu. I. 13) makes two right angles. Therefore (Eu. I. 14) KML will be in one continuous straight line, consequently a common perpendicular to the aforesaid straights AX, BX.

Quod erat etc.

COROLLARY II.

But again I am able hence to show that those two straights AX, BX, meeting with which the straight PFHD makes with the said AX, BX either two internal angles toward the same parts equal to two right angles, or

duobus rectis; aut consequenter (ex 13. et 15. primi) alternos sive externos, sive internos angulos inter se aequales; aut rursum, eodem titulo, externum (ut puta DHX) aequalem interno, et opposito HFX: quod, inquam, illae duae rectae neque ad infinitam earundem productionem coire inter se possint. Si enim ex quolibet puncto N ipsius AX demittatur ad BX perpendicularis NR, erit haec in ipsa hypothesi anguli acuti (quae utique sola obesse nobis posset) major (ex Cor. I. post 3. hujus) eo communi perpendiculo KL. Non igitur illae duae AX, BX, convenire unquam inter se poterunt.

Porro autem demonstratas hinc habes Propos. 27. et 28. Libri primi Euclidis; et quidem citra immediatam dependentiam a praecedentibus 16. et 17. ejusdem primi, cir-[48]ca quas oriri posset difficultas, quoties sub basi finita infinitilaterum esset triangulum; ad quale nempe triangulum provocare non dubitaret, qui eas duas AX, BX ad infinitam saltem distantiam inter se coituras censeret, quamvis anguli ad incidentem PFHD tales forent, quales supposuimus.

Praeterea, propter demonstratum commune perpendiculum KL, nequirent sane illae duae KX, LX ad suam partem punctorum X simul concurrere, quin etiam (ex facili intellecta superpositione) ad alteram etiam partem simul concurrerent reliquae et ipsae interminatae KA, LB. Quare duae rectae AX, BX clauderent spatium; quod est contra naturam lineae rectae.

Sed haec posteriora sunt. Nam in praecedentibus nusquam adhibui aut 16. aut 17. primi, nisi ubi clare ageretur de triangulo omni ex parte circumscripto, prout nempe in Proemio ad Lectorem ita me curaturum spoponderam.

consequently (from Eu. I. 13 and 15) alternate external or internal angles equal to one another, or again, from the same cause, an external (as suppose DHX) equal to an internal and opposite HFX; that, say I, those two straights not even in their infinite production can meet one another.

For if from any point N of AX is let fall to BX the perpendicular NR, this will be in the hypothesis of acute angle (which alone in any case can hinder us) greater (from P. III., Cor. I.) than the common perpendicular KL. Therefore those two straights AX, BX cannot ever meet one another.

But furthermore here you have Eu. I. 27 and 28 demonstrated, and indeed without immediate dependence from the preceding 16 and 17 of the same First Book, about [48] which difficulties could arise when the triangle should be of infinite sides on a finite base; to which sort of a triangle without doubt would refer one who believed that these two straights AX, BX met one another at least at an infinite distance, although the angles at the transversal PFHD were such as we have supposed.

Moreover, on account of the demonstrated common perpendicular KL, surely those two KX, LX cannot come together toward the part of the points X, since also (from a superposition easily understood) toward the other part also would meet at the same time the remaining and themselves unterminated KA, LB. Wherefore two straights AX, BX would enclose a space; which is contrary to the nature of the straight line.

But these things are later. For in the preceding I have never applied either Eu. I. 16 or 17, except where clearly it treats of a triangle bounded on every side, as indeed I promised I would so take care to do in the Preface to the Reader.

PROPOSITIO XXIV.

Iisdem manentibus: Dico quatuor simul angulos (fig. 27.) quadrilateri KDHK proximioris basi AB minores esse (in hypothesi anguli acuti) quatuor simul angulis quadrilateri KHLK remotioris ab eadem basi, atque ita quidem, sive illae duae AX, BX aliquando ad finitam distantiam incidant versus partes puncti X; sive nunquam inter se incidant; sed versus eas partes aut semper magis ad se invicem accedant, aut aliquando recipiant commune perpendiculum, post quod nempe (juxta Cor. II. praec. Propos.) ad easdem partes incipiant invicem dissilire.

Demonstratur. Verum hic supponimus portiones KK sumptas esse invicem aequales. Quoniam igitur (ex prae-[49]cedente) latus DK majus est latere HK, ac similiter HK majus latere LK; sumatur in HK portio MK aequalis ipsi LK, et in DK portio NK aequalis ipsi HK; junganturque MN, MK, LK; nimirum punctum K intermedium cum puncto L, et punctum K vicinius puncto B cum puncto M. Jam sic progredior. Quandoquidem latera trianguli KKL (initium semper ducam a puncto K viciniore puncto B) aequalia sunt lateribus trianguli KKM, et anguli comprehensi aequales, utpote recti; aequales etiam erunt (ex 4. primi) bases LK, MK; atque item aequales, qui correspondent invicem anguli, ad easdem

PROPOSITION XXIV.

The same remaining: I say the four angles together (fig. 27) of the quadrilateral KDHK nearer the base AB are less (in the hypothesis of acute angle) than the four angles together of the quadrilateral KHLK more remote from the same base; and indeed this is so, whether those two AX, BX somewhere at a finite distance meet toward the parts of the point X; or never meet one another; but toward those parts either ever more mutually approach each other, or somewhere receive a common perpendicular, after which of course (in accordance with Cor. II. of the preceding proposition) toward the same parts they begin mutually to separate.

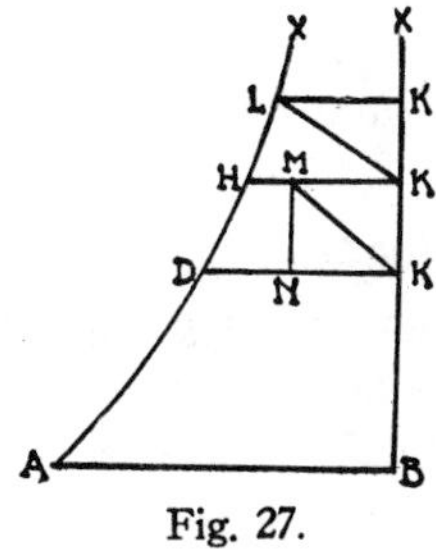

Fig. 27.

PROOF. Here however we suppose the portions KK assumed to be mutually equal. Since therefore (from the preceding) [49] the side DK is greater than the side HK, and similarly HK greater than the side LK, the portion MK in HK is assumed equal to LK, and in DK the portion NK equal to HK; and MN, MK, LK are joined, truly the intermediate point K with the point L, and the point K nearer to the point B with the point M.

Now I proceed thus.

Since indeed the sides of the triangle KKL (I make beginning always from the point K nearer to the point B) are equal to the sides of the triangle KKM, and the included angles equal, as being right, equal also will be (from Eu. I. 4) the bases LK, MK, and likewise equal the angles which correspond mutually, at these bases,

bases, nimirum angulus KLK angulo KMK, et angulus LKK angulo MKK. Igitur aequales etiam sunt residui NKM, et HKL. Quare, cum latera NK, KM, trianguli NKM aequalia itidem sint lateribus HK, KL trianguli HKL; aequales etiam erunt (ex eadem 4. primi) bases NM, HL; anguli KNM, KHL; ac tandem anguli KMN, KLH. Sunt autem in prioribus triangulis jam probati aequales anguli KLK, et KMK. Igitur totus angulus NMK aequalis est toti angulo HLK. Quare, cum omnes ad puncta K anguli sint recti, manifeste consequitur omnes simul quatuor angulos quadrilateri KNMK aequales esse omnibus simul quatuor angulis quadrilateri KHLK. Quoniam vero duo simul anguli ad puncta N, et M in quadrilatero KNMK majores sunt, in hypothesi anguli acuti, duobus simul angulis (ex Cor. post XVI. hujus) ad puncta D, et H in quadrilatero NDHM, seu quadrilatero KDHK; consequens inde est, ut (additis communibus rectis angulis ad puncta K) quatuor simul anguli quadrilateri KNMK, seu quadrilateri KHLK, majores sint (in hypothesi anguli acuti) quatuor simul angulis quadrilateri KDHK. Quod erat demonstrandum. [50]

COROLLARIUM.

Sed opportune observari hic debet, nihil defuturum factae argumentationi, quamvis angulus ad punctum L poneretur rectus, juxta hypothesin anguli acuti. Nam adhuc illa communis perpendicularis LK minor foret (ex Cor. I. post III. hujus) altera perpendiculari HK, ex qua propterea sumi adhuc posset portio MK aequalis prae-

indeed the angle KLK to the angle KMK, and the angle LKK to the angle MKK. Therefore equal also are the remainders NKM and HKL. Wherefore, since the sides NK, KM of the triangle NKM are equal in the same way to the sides HK, KL of the triangle HKL, equal also will be (from the same Eu. I. 4) the bases NM, HL, the angles KNM, KHL, and finally the angles KMN, KLH. But in the preceding triangles are already proved equal the angles KLK, KMK. Therefore the whole angle NMK is equal to the whole angle HLK.

Wherefore, since all angles at the points K are right, it follows manifestly all four angles together of the quadrilateral KNMK are equal to all four angles together of the quadrilateral KHLK.

But since the two angles together at the points N and M in the quadrilateral KNMK are greater, in hypothesis of acute angle, than the two angles together (from Cor. after P. XVI.) at the points D and H in the quadrilateral NDHM, or the quadrilateral KDHK, the consequence thence is, that (the common right angles at the points K being added) the four angles together of the quadrilateral KNMK, or the quadrilateral KHLK are greater (in hypothesis of acute angle) than the four angles together of the quadrilateral KDHK.

Quod erat demonstrandum. [50]

COROLLARY.

But it ought here opportunely to be observed, nothing will fail in the argument made, although the angle at the point L is assumed right, together with hypothesis of acute angle. For still that common perpendicular LK would be less (from Cor. I. to P. III.) than the other perpendicular HK, from which therefore still a portion MK could be assumed equal to the aforesaid LK.

dictae LK: Quo stante constat nullum posse obicem intercurrere.

SCHOLION.

Dubitari nihilominus posset, an ex quolibet puncto K (assumpto nimirum in BX ante occursum ipsius BX in alteram AX) perpendicularis educta versus partes rectae AX occurrere huic debeat (fig. 29.) in aliquo puncto L; dum nempe illae duae, ante praedictum occursum, ponantur ad se invicem semper magis accedere. Ego autem dico ita omnino secuturum.

Demonstratur. Assignatum sit in BX quodvis punctum K. Sumatur in AX quaedam AM aequalis summae ex ipsa BK, et dupla AB. Tum ex puncto M ducatur ad BX (juxta 12. primi) perpendicularis MN. Erit MN (juxta praesentem suppositionem) minor ipsa AB. Quare AM (facta aequalis summae ex ipsa BK, et dupla AB) major erit summa ipsarum BK, AB, et NM. Jam ostendere oportet eandem AM minorem esse summa ipsarum BN, AB, et MN, ut inde constet eam BN majorem esse praedicta BK, ac propterea punctum K jacere inter puncta B, et N. Jungatur BM. Erit latus AM (ex 20. primi) minus duobus simul reliquis lateribus AB, et BM. Rursum [51] latus BM (ex eadem 20. primi) minus erit duobus simul lateribus BN, et MN. Igitur latus AM multo minus erit tribus simul lateribus AB, BN, et NM. Hoc autem erat ostendendum, ut constaret punctum K jacere inter puncta B, et N. Inde autem consequens est, ut per-

Which standing, it follows that no hindrance can intervene.

SCHOLION.

Nevertheless it might be doubted, whether a perpendicular, from whatever point K (assumed indeed in BX before the meeting of this BX with the other AX) erected toward the parts of the straight AX, must meet this (fig. 29) in some point L; provided of course those two, before the aforesaid meeting, are assumed ever more to approach each other mutually.☼ But I say it will follow completely thus.

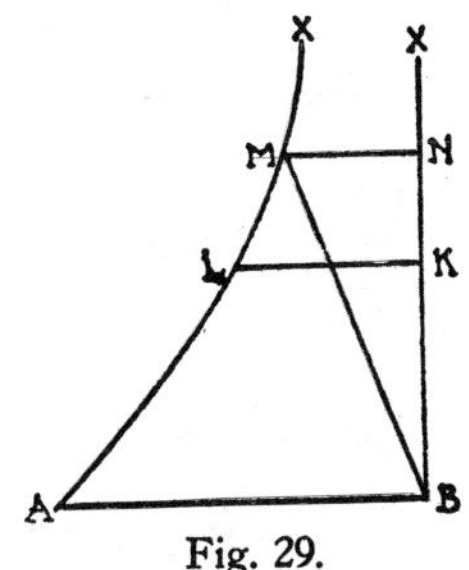

Fig. 29.

Proof. Let there be assigned in BX any point whatever K. In AX is taken a certain AM equal to the sum of this BK and of twice AB.

Then from the point M is drawn to BX (according to Eu. I. 12) the perpendicular MN. According to the present supposition, MN will be less than AB. Wherefore AM (made equal to the sum of BK and of double AB) will be greater than the sum of BK, AB, and NM. Now it behooves to show this same AM to be less than the sum of BN, AB, and MN, that thence it may follow this BN is greater than the aforesaid BK, and therefore the point K lies between the points B and N.

Join BM. The side AM will be (from Eu. I. 20) less than the two remaining sides together AB and BM. Again [51] the side BM (from the same Eu. I. 20) will be less than the two sides together BN and MN. Therefore the side AM will be by much less than the three sides together AB, BN, and NM. But this was to be shown, in order to deduce that the point K lies between the points B and N. Thence however it follows, that the perpen-

pendicularis ex puncto K educta versus partes ipsius AX occurrere huic debeat in aliquo puncto L inter puncta A, et M constituto; ne scilicet (contra 17. primi) secare debeat alterutram AB, aut MN perpendiculares eidem BX. Quod etc.

PROPOSITIO XXV.

Si duae rectae (fig. 30.) AX, BX in eodem plano existentes (una quidem sub angulo acuto in puncto A, et altera in puncto B perpendiculariter insistens ipsi AB) ita ad se invicem semper magis accedant versus partes punctorum X, ut nihilominus earundem distantia semper major sit assignata quadam longitudine, destruitur hypothesis anguli acuti.

Demonstratur. Assignata sit longitudo R. Si ergo in ea BX sumatur quaedam BK quantumlibet multiplex propositae longitudinis R; constat (ex praecedente Scholio) perpendicularem ex puncto K eductam versus partes ipsius AX in aliquo puncto L eidem occursuram; ac rursum (ex praesente hypothesi) constat eam KL majorem fore praedicta longitudine R. Porro intelligatur BK divisa in portiones KK, aequales singulas ipsi R, usque dum KB aequalis sit ipsi longitudini R. Tandem vero ex punctis K erectae sint ad BX perpendiculares occurrentes ipsi AX in punctis L, H, D, M, usque ad punctum N proximius puncto A. Jam sic progredior.

Erunt (ex Prop. praecedente) quatuor simul anguli quadrilateri KHLK, remotioris ab ea basi AB, majores

dicular from the point K erected toward the parts of AX must meet this in some point L stationed between the points A and M; else obviously (against Eu. I. 17) it must cut either AB or MN perpendiculars to BX.

Quod etc.

PROPOSITION XXV.

If two straights (fig. 30) AX, BX existing in the same plane (standing upon AB, one indeed at an acute angle in the point A, and the other perpendicular at the point B) so always approach more to each other mutually, toward the parts of the point X, that nevertheless their distance is always greater than a certain assigned length, the hypothesis of acute angle is destroyed.

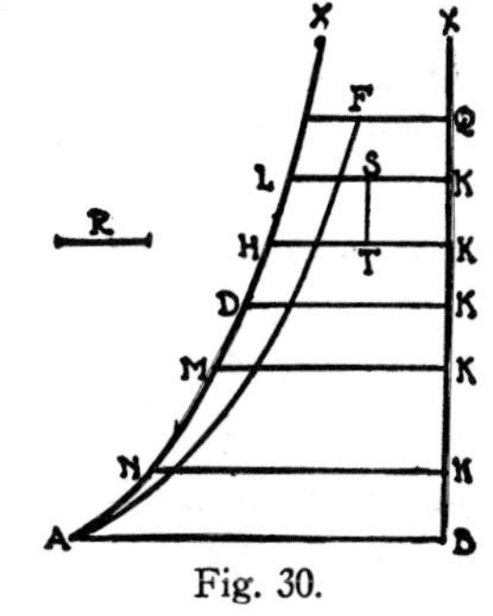

Fig. 30.

PROOF. Let R be the assigned length. If therefore in BX is assumed a certain BK any chosen multiple of the proposed length R; it follows (from the preceding scholion) that the perpendicular erected from the point K toward the parts of AX will meet it at some point L; and again (from the present hypothesis) it follows that this KL will be greater than the aforesaid length R. Furthermore BK is understood divided into portions KK, each equal to R, even until KB is itself equal to the length R. Finally from the points K are erected to BX perpendiculars meeting AX in points L, H, D, M, even to the point N nearest the point A.

Now I proceed thus.

The four angles together of the quadrilateral KHLK, more remote from the base AB, will be (from the pre-

[52] quatuor simul angulis quadrilateri KDHK, proximioris eidem basi; cujus itidem quadrilateri quatuor simul anguli majores erunt quatuor simul angulis subsequentis versus eandem basim quadrilateri KMDK. Atque ita semper usque ad ultimum quadrilaterum KNAB, cujus utique quatuor simul anguli minimi erunt, relate ad quatuor simul angulos singulorum ascendentium versus puncta X quadrilaterorum.

Quoniam vero tot aderunt praedicto modo recensita quadrilatera, quot sunt praeter basim AB demissae ex punctis ipsius AX ad rectam BX perpendiculares; expendenda est summa omnium simul angulorum, qui comprehenduntur in illis quadrilateris. Ponamus esse novem ejusmodi perpendiculares demissas, ac propterea novem itidem quadrilatera. Constat (ex 13. primi) aequales esse quatuor rectis angulos hinc inde comprehensos ad bina puncta illarum octo perpendicularium, quae mediae jaceant inter basim AB, et remotiorem perpendicularem LK. Itaque summa horum omnium angulorum erit 32 rectorum. Restant duo anguli ad perpendiculum LK, et duo ad basim AB. At anguli, unus quidem ad punctum K, et alter ad punctum B, supponuntur recti; angulus autem ad punctum L (ex Cor. post XXIII. hujus) est obtusus. Quapropter (etiam neglecto angulo acuto ad punctum A) summa omnium angulorum, qui comprehenduntur ab illis novem quadrilateris, excedet 35. rectos. Inde autem fit, ut quatuor simul anguli quadrilateri KHLK, remotioris a basi, minus deficiant a quatuor rectis, quam sit nona pars unius recti; et id quidem etiam si aequalis portio praedicta omnium angulorum summae contingeret singulis illis quadrilateris. Ergo minor adhuc erit insinuatus defectus, cum summa quatuor simul angu-

ceding proposition) greater [52] than the four angles together of the quadrilateral KDHK, nearer to this base; of which quadrilateral in the same way the four angles together will be greater than the four angles together of the quadrilateral KMDK subsequent toward this base. And so always even to the last quadrilateral KNAB, whose four angles together assuredly will be the least, in reference to the four angles together of each of the quadrilaterals ascending toward the points X.

But since are present as many quadrilaterals described in the aforesaid manner, as are, except the base AB, perpendiculars let fall from points of AX to the straight BX; the sum of all the angles together, which are comprehended in these quadrilaterals can be reckoned. We assume that there are nine such perpendiculars let fall, and therefore so nine quadrilaterals.

We get (from Eu. I. 13) as equal to four rights the angles comprehended hither and yon at the two points of those eight perpendiculars, which lie in the middle between the base AB and the more remote perpendicular LK. So the sum of all these angles will be 32 rights.

There remain two angles at the perpendicular LK, and two at the base AB. But the angles one indeed at the point K and the other at the point B are supposed right; but the angle at the point L (from the Cor. to P. XXIII.) is obtuse. Wherefore (even neglecting the acute angle at the point A) the sum of all the angles which are comprehended by these nine quadrilaterals exceeds 35 rights. But hence follows, that the four angles together of the quadrilateral KHLK, more remote from the base lack less from four rights than the ninth part of one right; and that indeed even if an equal portion of the aforesaid sum of all the angles pertained to each of those quadrilaterals.

Therefore less yet will be the occurring defect, since

lorum illius quadrilateri KHLK ostensa sit omnium maxima, relate ad [53] quatuor simul angulos reliquorum quadrilaterorum.

Sed rursum; juxta suppositionem, in qua procedit haec Propositio; assumi potest tanta longitudo ipsius BK, ut confici semper possint non tot quin plura quadrilatera sub basibus KK, aequalibus singulis illi assignatae longitudini R. Quare defectus quatuor simul angulorum illius remotioris quadrilateri KHLK a quatuor rectis ostendetur semper minor et una centesima, et una millesima, et sic sub quolibet assignabili numero una portiuncula unius recti.

Porro autem erunt semper (juxta praedictam suppositionem) ipsa LK, et HK majores designata longitudine R. Si ergo in KL, et KH sumantur KS, et KT aequales ipsi KK, seu longitudini R; erunt, juncta ST, duo simul anguli KST, KTS majores, in hypothesi anguli acuti, duobus simul angulis (ex Cor. post XVI. hujus) ad puncta H, et L in quadrilatero THLS, seu quadrilatero KHLK; ac propterea (additis communibus rectis angulis ad puncta K, K) erunt quatuor simul anguli quadrilateri KTSK majores quatuor simul angulis illius quadrilateri KHLK.

Jam vero: cum ex una parte stabile sit, ac datum quadrilaterum KTSK, utpote constans data basi KK, quae nimirum aequalis ponitur assignatae longitudini R, ac rursum constans duobus perpendiculis TK, SK eidem basi aequalibus, ac tandem jungente TS, quae evadit omnino determinata; et ex altera quatuor simul anguli stabilis illius, ac dati quadrilateri, ostensi jam sint majores quatuor simul angulis quadrilateri KHLK quantumlibet distantis ab ea basi AB: consequens utique fit, ut quatuor simul anguli stabilis illius, ac dati quadrilateri

the sum of the four angles together of this quadrilateral KHLK was shown the greatest of all, in relation to [53] the four angles together of the remaining quadrilaterals.

But again; in consequence of the supposition upon which this proposition proceeds, so great a length of BK can be assumed, that as many quadrilaterals as we choose may be made on bases KK, each equal to the assigned length R.

Wherefore the defect of the four angles together of this more remote quadrilateral KHLK from four rights is shown ever less both than a hundredth and than a thousandth, and thus under any assignable part of a right. Further however, LK and HK will be always (in accordance with the aforesaid supposition) greater than the designated length R. Therefore if in KL and KH are assumed KS and KT equal to KK or the length R; ST being joined, the two angles together KST, KTS will be greater, in hypothesis of acute angle, than the two angles together (from Cor. to P. XVI.) at the points H and L in the quadrilateral THLS, or the quadrilateral KHLK; and therefore (the common right angles at the points K, K being added) the four angles together of the quadrilateral KTSK will be greater than the four angles together of that quadrilateral KHLK.

But now, since on one hand is stable and given the quadrilateral KTSK, inasmuch as constant in the given base KK, which indeed is taken equal to the assigned length R, and again constant in the two perpendiculars TK, SK equal to this base, and finally in the joining TS, which comes out completely determinate; and on the other hand the four angles together of this stable and given quadrilateral have now been shown greater than the four angles together of the quadrilateral KHLK distant as far as we choose from the base AB; assuredly it follows that the four angles together of this stable and given

KTSK majores sint qualibet angulorum summa, quae quomodolibet deficiat a quatuor rectis; quandoquidem ostensum jam est designari semper posse tale aliquod quadrilaterum KHLK, [54] cujus quatuor simul anguli minus deficiant a quatuor rectis, quam sit quaevis designabilis unius recti portiuncula. Igitur quatuor simul anguli stabilis illius, ac dati quadrilateri, vel aequales sunt quatuor rectis, vel eisdem majores. Tunc autem (ex XVI. hujus) stabilitur hypothesis aut anguli recti, aut anguli obtusi; ac propterea (ex V. et VI. hujus) destruitur hypothesis anguli acuti.

Itaque constat destructum iri hypothesim anguli acuti, si duae rectae in eodem plano existentes ita ad se invicem semper magis accedant, ut nihilominus earundem distantia major semper sit assignata quadam longitudine. Hoc autem erat demonstrandum.

COROLLARIUM I.

At (destructa hypothesi anguli acuti) manifestum fit, ex 13. hujus, controversum Pronunciatum Euclidaeum; prout a me hoc loco declaratum iri spopondi in Scholio III. post XXI. hujus, ubi de conatu Nassaradini Arabis locuti sumus.

COROLLARIUM II.

Rursum ex hac Propositione, et ex praecedente XXIII. manifeste colligitur satis non esse ad stabiliendam Geometriam Euclidaeam duo puncta sequentia. Unum est: quod nomine parallelarum illas rectas censeamus, quae in eodem plano existentes commune aliquod obtinent perpendiculum. Alterum vero, quod omnes rectae in eodem plano existentes, quarum nullum commune sit per-

quadrilateral KTSK are greater than any sum of angles, which lacks however little you choose of being four right angles; since already it has been shown that a quadrilateral KHLK can always be designated such [54] that its four angles together shall fall short of four rights by less than any assignable part of a right. Therefore the four angles together of this stable and given quadrilateral either are equal to four rights or greater.

But then (from P. XVI.) is established the hypothesis either of right angle or of obtuse angle; and therefore (from Propp. V. and VI.) the hypothesis of acute angle is destroyed.

So is established that the hypothesis of acute angle will be destroyed, if two straights existing in the same plane so approach each other mutually ever more, that nevertheless their distance is always greater than any assigned length.

Hoc autem erat demonstrandum.

COROLLARY I.

But (the hypothesis of acute angle destroyed) the controverted Euclidean postulate is manifest from P. XIII.; just as in Scholion III. after P. XXI., where we spoke of the attempt of the Arab Nasiraddin, I promised would be disclosed by me in this place.

COROLLARY II.

On the other hand from this proposition, and from the preceding P. XXIII. is manifestly gathered that the two following points are not sufficient for establishing Euclidean geometry. One is: that we designate by the name of parallels those straights, which existing in the same plane possess a common perpendicular. The second indeed, that all straights existing in the same plane, of

pendiculum, ac propterea quae juxta assumptam Definitionem parallelae non sint, debeant ipsae in alterutram partem semper magis protractae inter se aliquando incidere, si non ad [55] finitam, saltem ad infinitam distantiam. Nam rursum demonstrare oporteret, quod duae quaelibet in eodem plano existentes, in quas recta quaepiam incidens duos ad easdem partes internos angulos efficiat minores duobus rectis, nusquam alibi possint ipsae recipere commune perpendiculum. Quod autem, hoc demonstrato, exactissime stabiliatur Geometria Euclidaea, infra constabit.

PROPOSITIO XXVI.

Si praedictae AX, BX (*fig.* 31.) *coire quidem inter se debeant, sed non nisi ad infinitam earundem productionem versus partes punctorum X: Dico nullum fore assignabile punctum I in ipsa AB, ex quo perpendicularis educta versus partes ipsius AX non occurrat ad finitam, seu terminatam distantiam eidem AX in aliquo puncto F.*

Demonstratur. Nam (ex praecedente hypothesi) unum aliquod erit in ipso AX punctum N, ex quo per-

which there is no common perpendicular, and therefore which according to the assumed definition are not parallel, must, being produced toward either part ever more, somewhere meet each other, if not at [55] a finite, at least at an infinite distance.

For again it would be requisite to demonstrate, that any two straights existing in the same plane, upon which a certain straight cutting makes two internal angles toward the same parts less than two right angles, nowhere else can receive a common perpendicular.

But that, this demonstrated, Euclidean geometry is most exactly established, will be shown below. ☼

PROPOSITION XXVI.

If the aforesaid AX, BX (fig. 31) must indeed meet each other, but only at their infinite production toward the parts of the point X: I say there will

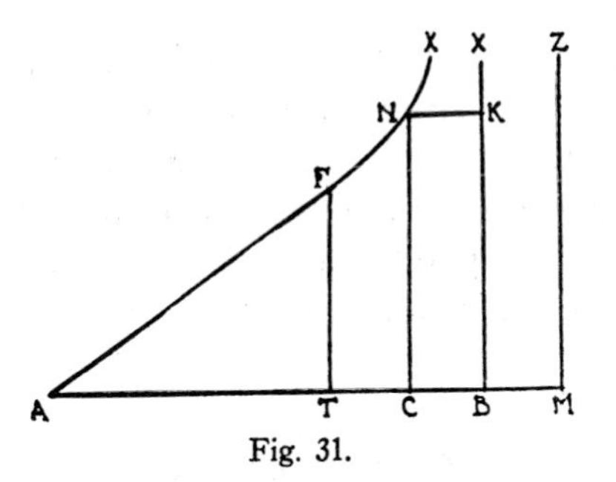

Fig. 31.

be no assignable point T in AB, from which a perpendicular erected toward the parts of AX does not at a finite, or terminated distance meet this AX in some point F.

Proof. For (from the preceding hypothesis) there will be in AX some point N, from which the perpen-

pendicularis NK demissa ad BX minor sit qualibet assignata longitudine, ut puta ea TB. Tum vero sumatur in TB portio CB aequalis ipsi NK, jungaturque CN. Constat angulum NCB acutum fore, in hypothesi anguli acuti. Ergo (ex 13. primi) obtusus erit, qui deinceps est angulus NCT. Igitur recta, quae ex puncto T (inter puncta A, et C constituto) perpendiculariter educatur versus partes ipsius AX, non incidet (ex 17. primi) in ullum punctum ipsius CN; ac propterea (ne claudat spatium cum AT, aut cum TC) occurret ipsi terminatae AN in aliquo puncto F. Igitur in ipsa etiam hypothesi anguli acuti (quam scimus obesse unice hic posse) nullum erit assignabile punctum T in ea AB, ex quo perpendiculariter educta versus partes ipsius AX non occurrat ad finitam, seu terminatam distantiam eidem AX in quodam puncto F. Quod etc. [56]

COROLLARIUM I.

Inde autem fit, ut assumpto in AB protracta quolibet puncto M, ex quo versus partes punctorum X educatur perpendicularis MZ, nequeat ipsa, etiamsi infinite producatur, occurrere praedictae AX; quia caeterum illa altera BX deberet (ex praemissa demonstratione) ad finitam distantiam occurrere eidem AX; quod est contra praesentem hypothesin.

COROLLARIUM II.

Ex quo rursum consequitur omnem perpendiculariter eductam ex quolibet puncto illius quantumlibet continuatae AB, sed non tamen infinite dissito, debere ad finitam distantiam occurrere praedictae AX; quatenus nempe supponatur omnem talem perpendiculariter eductam semper

dicular NK let fall to BX is less than any assigned length, as suppose than TB. But then is assumed in TB a portion CB equal to NK and CN is joined. In the hypothesis of acute angle, it is known that the angle NCB will be acute. Therefore (from Eu. I. 13) NCT, which is the adjacent angle, will be obtuse.

Therefore the straight, which is erected toward the parts of AX perpendicularly from the point T (disposed between the points A and C), does not meet (from Eu. I. 17) CN at any point; and therefore (lest it should inclose a space with AT, or with TC) it strikes the terminated AN in some point F.

Therefore even in the hypothesis of acute angle (which we know can here alone hinder) there will be in this AB no assignable point T, from which the perpendicular erected toward the parts of AX does not, at a finite or terminated distance, meet this AX in a certain point F. Quod erat etc. [56]

COROLLARY I.

But thence follows, that, point M being assumed in AB produced, from which is erected toward the parts of the points X a perpendicular MZ, this cannot, even if infinitely produced, meet the aforesaid AX; because otherwise that other straight BX must (from the foregoing demonstration) at a finite distance meet this AX; which is against the present hypothesis.

COROLLARY II.

From which again it follows, that every perpendicular, erected from any point (but not however infinitely removed) of this AB produced indefinitely, must at a finite distance meet the aforesaid AX, as soon as indeed it is assumed that every such perpendicular ever more,

magis, sine ullo certo limite accedere ad alteram semper continuatam AX.

COROLLARIUM III.

Unde tandem fit, ut ab illa AX neque ad infinitam ejusdem productionem secari possit ipsa BX; quia caeterum ex quodam illius AX ultra praedictam sectionem puncto intelligi posset demissa ad AB productam quaedam perpendicularis ZM; unde rursum fieret, ut ipsa BX (contra praesentem hypothesim) non ad infinitam, sed omnino ad finitam distantiam occurreret praedictae AX. Sed hoc postremum dictum sit ultra necessitatem. [57]

PROPOSITIO XXVII.

Si recta AX (*fig.* 32.) *sub aliquo, ut libet, parvo angulo educta ex puncto A ipsius AB, occurrere tandem debeat* (*saltem ad infinitam distantiam*) *cuivis perpendiculari BX, quae ad quandamlibet ab eo puncto A distantiam excitari intelligatur super ea incidente AB*: *Dico nullum jam fore locum hypothesi anguli acuti.*

Demonstratur. Ex quodam puncto K prope punctum A, ad libitum in ipsa AB designato, erigatur ad AB perpendicularis KL, quae utique (ex Cor. II. praecedentis Propositionis) occurret ipsi AX ad finitam, seu termina-

without any certain limit, approaches the other ever produced straight AX.

COROLLARY III.

Whence finally follows, that not even at its infinite production can BX be cut by that AX; because otherwise from any point of that AX beyond the aforesaid intersection a certain perpendicular ZM could be supposed let fall to AB produced; whence again would follow, that BX (against the present hypothesis) met the aforesaid AX not at an infinite, but wholly at a finite distance.

But this last dictum is beyond necessity. [57] ✲

PROPOSITION XXVII.

If a straight AX (fig. 32) drawn at any however small angle from the point A of AB, must at length meet (anyhow at an infinite distance) any perpendicular BX, which is supposed erected at any distance from this point A upon the secant AB: I say there will then be no more place for the hypothesis of acute angle.

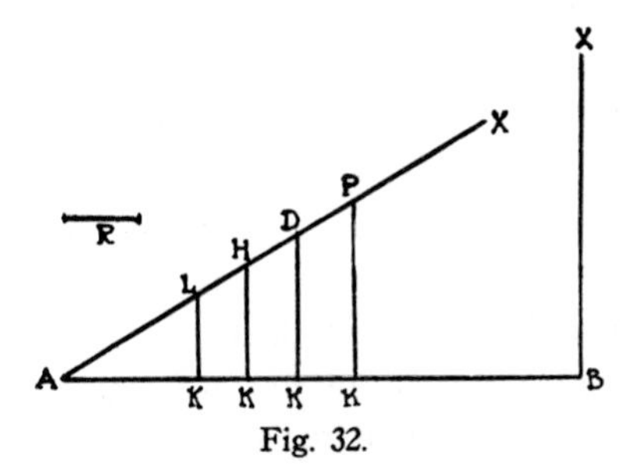

Fig. 32.

Proof. From any point K chosen at will in AB near the point A, the perpendicular KL is erected to AB, which certainly (from Cor. II. of the preceding proposition) meets AX at a finite or terminated distance in

tam distantiam in aliquo puncto L. Jam vero constat sumi posse in KB portiones KK aequales singulas cuidem assignabili longitudini R, et eas plures quolibet assignabili numero finito; quandoquidem punctum B statui potest; juxta praesentem suppositionem; in quantalibet distantia ab eo puncto A. Itaque ex aliis punctis K erigantur ad AB perpendiculares KH, KD, KP, quae omnes (ex praecitato Corollario) occurrent rectae AX in quibusdam punctis H, D, P; atque ita circa reliqua puncta K uniformiter designata versus punctum B. Constat secundo (ex 16. primi) angulos ad puncta L, H, D, P, fore omnes obtusos versus partes punctorum X; atque item (ex 13. ejusdem primi) angulos ad praedicta puncta fore omnes acutos versus punctum A. Igitur (ex Cor. II. post 3. hujus) latus KH majus erit latere KL; latus KD majus latere KH; atque ita semper, procedendo versus puncta X. Constat tertio quatuor simul angulos quadrilateri KLHK majores fore quatuor simul angulis quadrilateri KHDK: nam id in simili demonstratum jam est in XXIV. hujus. Constat quarto idem similiter valere de quadrilatero KHDK relate ad quadrilaterum KDPK; atque ita semper, procedendo ad qua-[58]drilatera remotiora ab eo puncto A.

Quoniam igitur tot aderunt (ut in XXV. hujus) praedicto modo recensita quadrilatera, quot sunt, praeter primam LK, demissae ex punctis ipsius AX perpendiculares ad rectam AB; constabit uniformiter (si ponamus novem, praeter primam, demissas ejusmodi perpendiculares) summam omnium angulorum, qui comprehenduntur ab illis novem quadrilateris, excedere 35. rectos; ac

some point L. But now it holds that there may be assumed in KB portions KK each equal to a certain assignable length R, and these more than any assignable finite number; since indeed the point B can be situated, in accordance with the present supposition, at however great a distance from this point A.

And accordingly from the other points K are erected to AB perpendiculars KH, KD, KP, which all (from the aforesaid corollary) meet the straight AX in certain points H, D, P; and so about the remaining points K uniformly designated toward the point B.

It holds secondly (from Eu. I. 16) that the angles at the points L, H, D, P will all be obtuse toward the parts of the points X; and just so (from Eu. I. 13) the angles at the aforesaid points will all be acute toward the point A.

Therefore (from Cor. II. to P. III.) the side KH will be greater than the side KL; the side KD greater than the side KH; and so always proceeding toward the points X.

It holds thirdly that the four angles together of the quadrilateral KLHK will be greater than the four angles together of the quadrilateral KHDK: for this in like case has already been demonstrated in P. XXIV.

It holds fourthly that the same is valid likewise of the quadrilateral KHDK in relation to the quadrilateral KDPK; and so on always, proceeding to quadrilaterals [58] more remote from this point A.

Since therefore are present (as in P. XXV.) as many quadrilaterals described in the aforesaid mode, as there are, except the first LK, perpendiculars let fall from points of AX to the straight AB, it will hold uniformly (if we assume nine perpendiculars of this sort let fall, besides the first) the sum of all the angles which are comprehended by these nine quadrilaterals will exceed 35 right

propterea quatuor simul angulos primi quadrilateri KLHK, quod quidem in hac ratione ostensum est omnium maximum, minus deficere a quatuor rectis, quam sit nona pars unius recti. Quare; multiplicatis ultra quemlibet assignabilem finitum numerum eisdem quadrilateris, procedendo semper versus partes punctorum X; constabit similiter (ut in eadem praecitata) quatuor simul angulos stabilis illius quadrilateri KHLK minus deficere a quatuor rectis, quam sit quaelibet assignabilis unius recti portiuncula. Igitur quatuor simul illi anguli vel aequales erunt quatuor rectis, vel eisdem majores. Tunc autem (ex XVI. hujus) stabilitur hypothesis aut anguli recti, aut anguli obtusi; ac propterea (ex V. et VI. hujus) destruitur hypothesis anguli acuti.

Itaque constat nullum jam fore locum hypothesi anguli acuti, si recta AX sub aliquo, ut libet, parvo angulo, educta ex puncto A ipsius AB occurrere tandem debeat (saltem ad infinitam distantiam) cuivis perpendiculari BX, quae ad quantamlibet ab eo puncto A distantiam excitari intelligatur super ea incidente AB. Quod erat etc.

SCHOLION I.

Et hoc est, quod praedixi in Cor. II. post XXV. hujus; nullum scilicet superfuturum locum hypothesi an-[59]guli acuti, seu stabilitum exactissime iri Geometriam Euclidaeam; si duae quaelibet in eodem plano existentes rectae, ut puta AX, BX, in quas incidens recta AB (sumpto puncto B in quantalibet distantia a puncto A) duos cum eisdem ad easdem partes punctorum X angulos efficiat minores duobus rectis; si (inquam) nusquam alibi (hoc

angles; and therefore the four angles together of the first quadrilateral KLHK, which indeed in this regard has been shown the greatest of all, will fall short of four right angles by less than the ninth part of one right angle. Wherefore, these quadrilaterals being multiplied beyond any assignable finite number, proceeding always toward the parts of the points X, it holds in the same way (as in the same already recited theorem) that the four angles together of this stable quadrilateral KHLK will fall short of four right angles less than any assignable little portion of one right angle.

Therefore these four angles together will be either equal to four right angles, or greater.

But then (from P. XVI.) is established the hypothesis of right angle or of obtuse angle; and therefore (from Propp. V. and VI.) is destroyed the hypothesis of acute angle.

So then it holds, that there will be no place for the hypothesis of acute angle, if the straight AX drawn under however small angle from the point A of AB must at length meet (anyhow at an infinite distance) any perpendicular BX, which is supposed erected at any distance from this point A upon this secant AB.

Quod erat etc.

SCHOLION I.

And this it is, that I said before in Cor. II. to P. XXV.; obviously that no place would remain over for the hypothesis of acute angle, [59] or Euclidean geometry would be most exactly established, if any two straights existing in the same plane, as suppose AX, BX, which the straight AB meeting (the point B being assumed at a distance from the point A as great as you choose) makes with them toward the same parts of the points X two angles less than two right angles, if (I say) nowhere

stante) possint illae recipere commune perpendiculum. Tunc enim illae duae AX, BX, semper magis ad se invicem accedent; nimirum vel intra quendam determinatum limitem, prout in XXV. hujus; vel sine ullo certo limite, ac propterea usque ad occursum saltem post infinitam productionem, prout in hac XXVII. Constat autem in utroque praedictorum casuum ostensam jam esse destructionem hypothesis anguli acuti. Quod intendebatur.

SCHOLION II.

Atque id rursum est, quod spopondi in fine Scholii IV. post XXI. hujus, prout ex ipsis terminis clare elucescit.

SCHOLION III.

Praeterea observari hic velim discrimen inter hanc Propos. et praecedentem XVII. Nam ibi (recole fig. 15.) ostensa est destructio hypothesis anguli acuti, si (existente, ut libet parva, recta AB) omnis BD sub quovis acuto angulo educta, occurrere tandem debeat in quodam puncto K ipsi perpendiculari AH productae. Hic autem (viceversa) permittitur quidem designatio cujusvis parvissimi acuti anguli ad punctum A, dum tamen interjecta AB, ad quam erigenda est perpendicularis indefinita [60] BX, statui possit quantaelibet longitudinis.

PROPOSITIO XXVIII.

Si duae rectae AX, BX (quarum prior sub angulo acuto, et altera ad perpendiculum eductae sint versus easdem partes ex quantalibet recta AB) semper magis

at another place (this standing) they can admit a common perpendicular.

For then these two AX, BX mutually approach each other ever more, indeed either within a certain determinate limit, as in P. XXV., or without any certain limit, and therefore even to meeting, anyhow after infinite production, as in P. XXVII.

But it holds that in either of the aforesaid cases the destruction of the hypothesis of acute angle has now been shown.

Quod intendebatur.

SCHOLION II.

And again this it is, that I promised at the end of Scholion IV. after P. XXI., as from the very terms clearly appears.

SCHOLION III.

Moreover I could wish here to be observed the difference between this proposition and the preceding P. XVII. For there (recall fig. 15) has been shown the destruction of the hypothesis of acute angle, if (the straight AB being as small as you choose) every BD erected at whatever acute angle, must at length meet in some point K the perpendicular AH produced.

But here (*vice versa*) in fact is permitted the designation of however most small an acute angle at the point A while still the sect AB to which is to be erected the indefinite perpendicular [60] BX, may be taken of any length whatever.

PROPOSITION XXVIII.

If two straights AX, BX (produced from any-sized straight AB toward the same parts, the first under an acute angle, and the other perpendicularly) mu-

sine ullo certo limite ad se invicem accedant, praeterquam ad infinitam earundem productionem; Dico omnes angulos (fig. 33.) ad quaelibet puncta L, H, D ipsius AX, ex quibus demittantur ad rectam BX perpendiculares LK, HK, DK; tum fore omnes obtusos versus partes puncti A; tum fore semper minores, qui magis distant ab eo puncto A; ac tandem angulos magis, ac magis distantes ab eodem puncto A, semper magis sine ullo certo limite accedere ad aequalitatem cum angulo recto.

Demonstratur. Et prima quidem pars constat ex Cor. I. post XXIII. hujus. Secunda vero pars ita evincitur. Nam duo simul anguli ad LK versus basim AB majores sunt (ex Cor. post XVI. hujus) duobus simul internis, et oppositis angulis ad HK versus eandem basim AB. Sunt autem inter se aequales, utpote recti, anguli ad utrunque punctum K versus basim AB. Ergo angulus obtusus ad L versus basim AB major est angulo obtuso ad H versus eandem basim AB. Simili modo ostendetur praedictum angulum obtusum ad H majorem esse angulo obtuso ad punctum D. Atque ita semper, procedendo versus puncta X.

Tertia tandem pars majore indiget disquisitione. Si ergo fieri potest, assignatus sit (fig. 34.) quidam angulus MNC, quo semper major sit, aut saltem non minor, excessus cujusvis ex praedictis angulis obtusis supra angulum rectum. Constat (ex XXI. hujus) latera NM, NC comprehendentia illum angulum MNC taliter produci posse, ut perpendicularis MC, ex quodam puncto M ipsius MN [61] demissa ad NC, major sit (in ipsa etiam hypo-

tually approach each other ever more without any certain limit, save at their infinite production; I say all angles (fig. 33) at any points L, H, D of AX, from which are let fall to the straight BX perpendiculars LK, HK, DK, first will all be obtuse toward the parts of the point A, secondly will be ever less, the more distant from this point A, and finally the angles more and more distant from this same point A ever more without any certain limit approach to equality with a right angle.

PROOF. The first part follows indeed from Cor. I. to P. XXIII. The second part however is proved thus. For the two angles together at LK toward the base AB are greater (from Cor. to P. XVI.) than the two internal and opposite angles together at HK toward the same base AB.

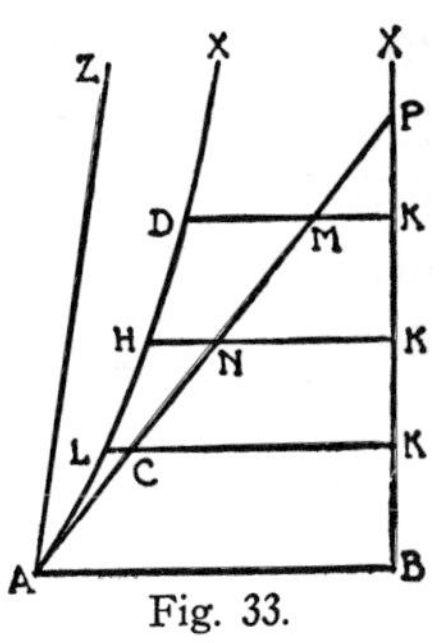

Fig. 33.

But the angles at each point K toward the base AB are equal to each other, as being right. Therefore the obtuse angle at L toward the base AB is greater than the obtuse angle at H toward the same base AB.

In like manner is shown that the aforesaid obtuse angle at H is greater than the obtuse angle at the point D.

And thus ever, proceeding toward the points X.

Finally the third part requires a longer disquisition. If therefore it can be done, let there be assigned (fig. 34) a certain angle MNC, than which is always greater, or anyhow not less, the excess of any of the aforesaid obtuse angles above a right angle. It follows (from P. XXI.) that the sides NM, NC comprehending that angle MNC can be so produced that the perpendicular MC from a certain point M of MN [61] let fall upon NC may be

thesi anguli acuti) qualibet finita assignata longitudine, ut puta praedicta basi AB. Hoc stante: assumatur in BX (fig. 35.) quaedam BT aequalis ipsi CN; educaturque ex puncto T versus AX perpendicularis TS, quae nempe (ex Scholio post XXIV. hujus) occurret ipsi AX in quodam puncto S. Deinde ex puncto S demittatur ad AB perpendicularis SQ. Cadet haec (propter 17. primi) ad partes anguli acuti SAB inter puncta A, et B. Porro acutus erit angulus QST in quadrilatero QSTB, cum reliqui tres anguli sint recti; ne (contra V. et VI. hujus) incidamus in hypothesin aut anguli recti, aut anguli obtusi. Hinc recta SQ major erit (ex Cor. I. post 3. hujus) recta BT, sive CN; ac rursum angulus ASQ major erit excessu, quo angulus obtusus AST excedit angulum rectum, et sic major angulo MNC. Ducatur igitur quaedam SF secans AQ in F, et efficiens cum SA angulum aequalem ipsi MNC. Deinde ex puncto A ducatur ad SF productam perpendicularis AO. Cadet punctum O (ex 17. primi) infra punctum F, cum angulus AFS (ex 16. ejusdem primi) sit obtusus. Tandem vero; cum FS major

greater (even in the hypothesis of acute angle) than any assigned finite length, as for instance the aforesaid base AB.

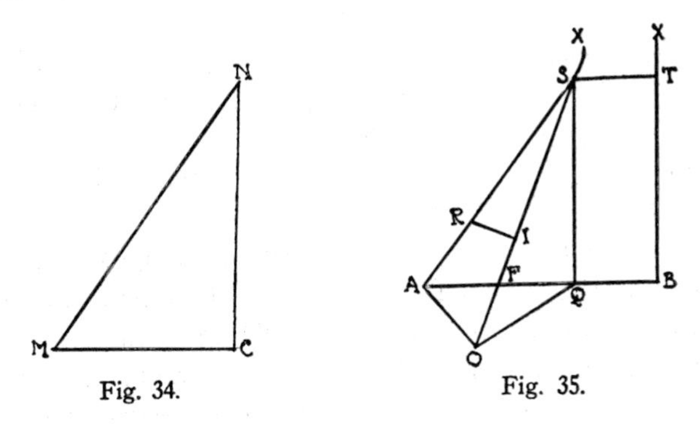

Fig. 34.

Fig. 35.

This standing; assume in BX (fig. 35) a certain BT equal to CN, and erect from the point T toward AX the perpendicular TS, which obviously (from Scholion after P. XXIV.) meets AX in a certain point S. Then from the point S let fall to AB the perpendicular SQ.

This falls (because of Eu. I. 17) toward the parts of the acute angle SAB between the points A and B. Again, acute will be the angle QST in the quadrilateral QSTB, since the remaining three angles are right; else (against Propp. V. and VI.) we come upon the hypothesis either of right angle or of obtuse angle.

Hence the straight SQ will be greater (from Cor. I. to P. III.) than the straight BT, or CN; and again the angle ASQ will be greater than the excess by which the obtuse angle AST exceeds a right angle, and thus greater than the angle MNC. Draw therefore a certain SF cutting AQ in F and making with SA an angle equal to MNC. Then from the point A draw to SF produced the perpendicular AO. The point O falls (from Eu. I. 17) below the point F, since the angle AFS (by Eu. I. 16) is obtuse.

sit (ex 19. primi) ipsa QS, et sic multo major ipsa BT, sive CN; sumatur in FS portio IS aequalis ipsi CN, et ex puncto I erigatur ad FS perpendicularis IR occurrens in puncto R ipsi AS. Cadet autem punctum R inter puncta A, et S: si enim caderet in aliquod punctum ipsius AF, haberemus in eodem triangulo (contra 17. primi) duos angulos majores duobus rectis, cum angulus ad punctum F versus partes puncti A ostensus jam sit obtusus.

Post tantum apparatum sic concludo. Quandoquidem in quadrilatero AOIR recti sunt anguli ad puncta O, et I; et est acutus angulus (ex 17. primi) ad punctum A, propter rectum angulum AOS; ac rursum est obtusus (ex 16. [62] ejusdem primi) angulus IRA, cum rectus sit angulus RIS: consequens tandem est (ex Cor. II. post 3. hujus) ut latus AO majus sit latere IR. At (juncta OQ) latus AQ majus est (ex 18. primi) latere AO propter angulum obtusum in O, cum angulus AOS factus sit rectus. Igitur recta AQ multo major erit recta IR, sive (ex 26. primi) recta MC, et sic multo major recta AB, pars toto; quod est absurdum.

Non igitur ullus assignari potest angulus MNC, quo semper major sit, aut saltem non minor excessus cujusvis ex praedictis angulis obtusis supra angulum rectum. Quare anguli illi obtusi, magis ac magis distantes ab eo puncto A, semper magis sine ullo certo limite accedent ad aequalitatem cum angulo recto. Quod erat postremo loco demonstrandum.

Finally, however; since FS is greater (by Eu. I. 19) than QS and so much greater than BT or CN, assume in FS the piece IS equal to CN, and from the point I erect to FS the perpendicular IR meeting AS in the point R.

But the point R falls between the points A and S: for if it fell on any point of AF, we would have in the same triangle (against Eu. I. 17) two angles greater than two right angles, since the angle at the point F toward the parts of the point A has already been shown obtuse.

After so much preparation thus I conclude. Since in the quadrilateral AOIR the angles at the points O and I are right, and the angle at the point A (by Eu. I. 17) is acute because of the right angle AOS, and again the angle IRA (by Eu. I. 16) is obtuse, [62] since the angle RIS is right: the consequence finally is (by Cor. II. to P. III.) that the side AO is greater than the side IR.

But (OQ joined) the side AQ is greater (by Eu. I. 19) than the side AO, because of the obtuse angle at O, since the angle AOS was made right.

Therefore the straight AQ will be much greater than the straight IR, or (by Eu. I. 26) than the straight MC, and so much greater than the straight AB, the part than the whole; which is absurd.

Therefore it is not possible to assign any one angle MNC, than which always is greater, or anyhow not less, the excess of each of the aforesaid obtuse angles above a right angle.

Wherefore those obtuse angles, more and more distant from this point A, ever more without any certain limit approach to equality with a right angle.

Quod erat postremo loco demonstrandum.

COROLLARIUM.

Hoc autem stante, quod postremo loco demonstratum est, manifeste consequitur, duas illas AX, BX, in infinitum protractas, commune tandem habituras, vel in duobus distinctis punctis, vel in uno, eodemque puncto X infinite dissito, perpendiculum. Rursum vero, quod non in duobus distinctis punctis haberi possit commune istud perpendiculum, ex eo manifeste liquet; quia caeterum (ex Cor. II. post XXIII. hujus) inciperent inde illae rectae invicem dissilire, et sic neque ad infinitam distantiam inter se concurrerent; quin etiam (contra expressam suppositionem) non ad se invicem, sine ullo certo limite, semper magis versus eas partes accederent. Itaque in uno, eodemque puncto X infinite dissito commune haberent perpendiculum. [63]

PROPOSITIO XXIX.

Resumpta fig. (33.) *praecedentis Propositionis: Dico omnem rectam AC, quae secet angulum BAX, aliquando ad finitam, seu terminatam distantiam (etiam in hypothesi anguli acuti) occursuram ipsi BX in quodam puncto P, dum nempe illa AC semper magis protrahatur versus partes punctorum X.*

Demonstratur. Et primo quidem (ne recta AC spatium claudat cum ea AX) occurret ipsa ad finitam distantiam rectis LK, HK, DK in quibusdam punctis C, N, M; occurret, inquam, nisi antea (ad finitam utique distantiam, prout intendimus) occurrat ipsi BX in aliquo puncto inter punctum B, et unum aliquod punctorum K constituto. Deinde (ex Cor. I. post XXIII. hujus) obtusi erunt anguli ACK, ANK, AMK. Praeterea anguli isti, semper obtusi, accedent (ex praecedente) sine ullo

COROLLARY.

But this standing, which in the last case was demonstrated, it manifestly follows that those straights AX, BX, produced infinitely will finally have, either in two distinct points, or in one same point X infinitely distant, a common perpendicular.

But again, that this common perpendicular cannot be had in two distinct points flows manifestly from this, because otherwise (by Cor. II. to P. XXIII.) those straights would thence begin mutually to separate, and so not meet each other at an infinite distance; so that also (against the express supposition) they would not mutually approach each other without any certain limit ever more toward those parts.

So they must have the common perpendicular in one same point X infinitely distant. [63]

PROPOSITION XXIX.

Resuming fig. 33 of the preceding proposition: I say every straight AC, which cuts angle BAX, finally at a finite, or terminated distance (even in the hypothesis of acute angle) will meet BX in a certain point P, if only AC be produced ever more toward the parts of the points X.

Proof. And first indeed (lest straight AC include space with AX) it must meet at finite distance the straights LK, HK, DK in certain points C, N, M; must meet, I say, unless before (and that at a finite distance, just as we maintain) it meets BX in some point between the point B and one of the points K.

Then (from Cor. I. to P. XXIII.) the angles ACK, ANK, AMK will be obtuse.

Moreover those angles, always obtuse, approach

certo limite ad aequalitatem cum angulo recto, quoties nempe illa AC non nisi ad infinitam distantiam occursura putetur ipsi BX. Igitur deveniri posset ad talem ordinatam KMD, ad quam angulus AMK minus superaret angulum rectum, quam sit ille angulus DAC. Tunc autem angulus DAC, sive DAM, una cum angulo AMD major erit uno recto. Quare; addito obtuso angulo ADM; tres simul anguli trianguli ADM majores erunt duobus rectis, quod est contra hypothesin anguli acuti. Igitur omnis recta AC, quae secet illum angulum BAX, aliquando ad finitam, seu terminatam distantiam (etiam in hypothesi anguli acuti) occurret ipsi BX in quodam puncto P. Quod etc. [64]

COROLLARIUM I.

Hinc nulla AZ, quae versus partes punctorum X angulum acutum efficiat majorem illo BAX, occurrere unquam poterit, sive ad finitam sive ad infinitam distantiam ipsi BX. Quatenus enim ita contingeret, jam illa AX, dividens angulum BAZ, deberet (contra praemissam suppositionem) ad finitam distantiam occurrere ipsi BX; prout demonstratum id est de recta AC dividente angulum BAX.

COROLLARIUM II.

Praeterea sequitur nullum fore determinatum acutum angulum omnium maximum, sub quo educta ex puncto A ad finitam distantiam occurrat illi BX. Si enim versus partes puncti X punctum quodvis assumas, quod sit altius puncto P, constat rectam jungentem punctum A

(from the preceding proposition) without any certain limit, to equality with a right angle, when indeed that AC is supposed to meet BX only at an infinite distance.

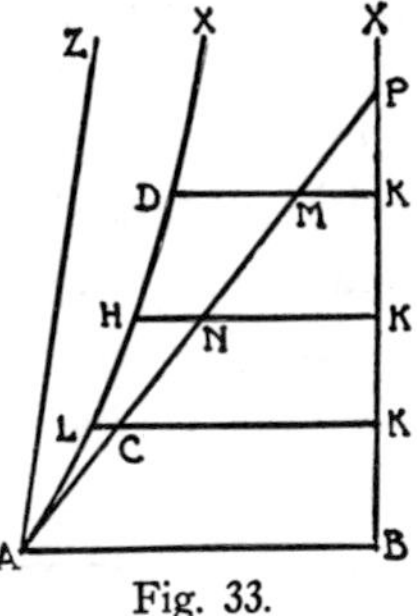

Fig. 33.

Therefore such an ordinate KMD can be reached that at it the angle AMK exceeds a right angle by less than the angle DAC. But then angle DAC, or DAM, together with angle AMD will be greater than a right angle.

Wherefore the obtuse angle ADM being added, the three angles together of the triangle ADM will be greater than two right angles, which is against the hypothesis of acute angle.

Therefore every straight AC, which cuts that angle BAX, finally at a finite or terminated distance (even in the hypothesis of acute angle) must meet BX in a certain point P.

Quod etc. [64]

COROLLARY I.

Hence no straight AZ, which toward the parts of the points X makes an acute angle greater than BAX can ever meet BX, either at a finite or at an infinite distance.

For as far as so should happen, now AX, dividing angle BAZ, ought (against the premised supposition) to meet BX at a finite distance, as this is demonstrated of the straight AC dividing angle BAX.

COROLLARY II.

Moreover it follows that no determinate acute angle will be the maximum of all under which a straight line produced from point A meets BX at finite distance.

For if toward the parts of the point X you assume any point higher than point P, it follows that the straight

cum illo puncto altiore majorem angulum effecturam cum ipsa AB, quam sit angulus BAP. Atque ita semper sine ullo termino intrinseco. Quare angulus BAX (dum scilicet ipsa AX, et semper accedat ad eam BX, et non nisi ad infinitam distantiam in eandem incidat) erit limes extrinsecus acutorum omnium angulorum, sub quibus rectae eductae ex illo puncto A ad finitam distantiam occurrunt praedictae BX.

PROPOSITIO XXX.

Cuivis terminatae AB insistat ad perpendiculum (fig. 36.) quaedam indefinita BX. Dico primo rectam AY, perpendiculariter elevatam versus partes easdem super illa AB, fore limitem unum intrinsecum earum omnium, quae ex illo puncto [65] *A versus easdem partes eductae commune aliquod (juxta hypothesin anguli acuti) in duobus distinctis punctis obtinent perpendiculum cum altera indefinita BX. Dico secundo nullum fore acutum angulum omnium minimum, sub quo educta ex praedicto puncto A commune aliquod (juxta praedictam hypothesin) in duobus distinctis punctis obtineat perpendiculum cum eadem BX.*

Demonstratur prima pars. Quoniam enim illa AY commune obtinet cum altera BX perpendiculum AB in

joining point A with this higher point will make with AB a greater angle than angle BAP.

And so ever without any intrinsic end. ✡

Wherefore angle BAX (since indeed AX both always approaches to BX, and meets it only at an infinite distance) will be the outside limit of all acute angles under which straights produced from that point A meet the aforesaid BX at a finite distance.

PROPOSITION XXX.

To any terminated straight AB stands at right angles (fig. 36) *a certain unbounded straight BX. I say firstly, that the straight AY, erected perpendicularly toward the same parts upon AB, will be one intrinsic limit of all those straights, which drawn from the*

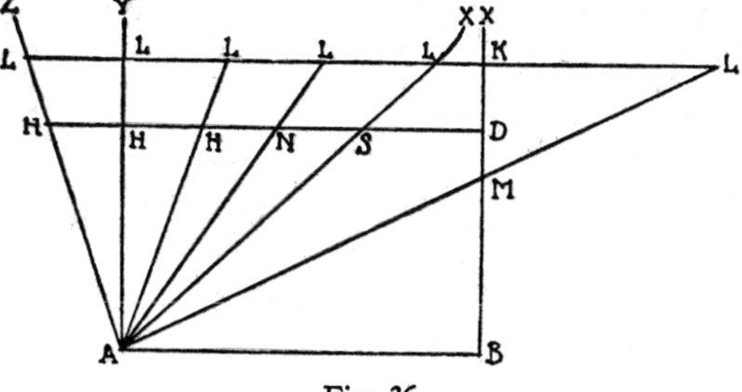

Fig. 36.

point [65] *A out toward the same parts have (in the hypothesis of acute angle) a common perpendicular in two distinct points with the other unbounded straight BX.* ✡ *I say secondly, that no acute angle will be the minimum of all, produced under which a straight from the aforesaid point A (in the aforesaid hypothesis) has in two distinct points a common perpendicular with BX.*

Proof of the First Part. For since AY has in common at two distinct points A and B the perpendicular

duobus distinctis punctis A, et B; si educatur versus easdem partes sub angulo obtuso quaepiam AZ, constat nullum ad eas partes esse posse in duobus distinctis punctis commune perpendiculum ipsarum AZ, BX; ne scilicet ex consecuturo quadrilatero continente quatuor angulos majores quatuor rectis incidamus (ex XVI. hujus) in hypothesin jam reprobatam anguli obtusi, contra suppositam hoc loco hypothesin anguli acuti. Igitur illa perpendicularis AY erit ex ista parte limes intrinsecus earum omnium, quae ex illo puncto A versus easdem partes eductae commune aliquod (juxta illam hypothesin anguli acuti) in duobus distinctis punctis obtineant perpendiculum cum altera indefinita BX. Quod erat primum.

Demonstratur secunda pars. Si enim fieri potest; esto quidam angulus acutus omnium minimus, sub quo educta AN commune habeat cum illa BX in duobus distinctis punctis perpendiculum ND. Tum assumpto in BX altiore puncto K, ex eo educatur ad BX perpendicularis KL, ad quam ex puncto A demittatur (juxta 12. primi) perpendicularis AL. Jam vero, si haec AL occurrat in quodam puncto S ipsi ND, constat sane angulum BAL minorem fore eo BAN, qui propterea non erit omnium minimus, sub quo educta AN commune habeat cum illa BX in duobus distinctis punctis perpendiculum ND. [66] Porro autem ab ea perpendiculari AL secari praedictam ND in quodam ejus intermedio puncto S sic demonstratur.

Et primo quidem non posse ab ea AL secari ipsam BK in quodam puncto M constare absolute potest ex 17. primi, ne scilicet in eodem triangulo MKL duos habeamus angulos rectos in punctis K, et L; praeterquam quod in hoc ipso haberemus intentum contra illum angulum BAN, ne scilicet in hac tali ratione censeatur omnium

AB with BX; if any straight AZ is drawn toward the same parts under an obtuse angle, it follows there can be toward these parts in two distinct points no common perpendicular to AZ, BX. Otherwise from the resulting quadrilateral containing four angles greater than four right angles, we hit (from P. XVI.) upon the already rejected hypothesis of obtuse angle, against the hypothesis of acute angle in this place assumed.

Therefore that perpendicular AY will be from that side an intrinsic limit of all the straights which drawn from the point A toward the same parts have (in the hypothesis of acute angle) at two distinct points a common perpendicular with the other unbounded straight BX

Quod erat primum.

PROOF OF THE SECOND PART. For if it were possible, let a certain acute angle be the least of all, drawn under which AN has with BX in two distinct points the common perpendicular ND. Then in BX a higher point K being assumed, from this erect to BX the perpendicular KL, upon which from the point A let fall (by Eu. I. 12) the perpendicular AL.

But now, if this AL meets ND in any point S, it certainly follows that angle BAL will be less than BAN, which therefore will not be the least of all drawn under which AN has with BX in two distinct points a common perpendicular ND. [66]

But furthermore that the aforesaid perpendicular ND is cut by this perpendicular AL in some intermediate point of it S is thus demonstrated.

And first indeed, that BK cannot be cut by AL in any point M follows absolutely from Eu. I. 17, since otherwise in the same triangle MKL we would have two right angles at the points K and L, apart from the fact that in this case we would have our assertion about that angle BAN, that it is not in such circumstances the least of all.

minimus. Rursum vero nequit AL esse continuatio ipsius AN; quia caeterum in quadrilatero NDKL quatuor haberemus angulos rectos, contra hypothesim anguli acuti. Sed neque eam DN protractam secare potest in quovis ulteriore puncto H; quia angulus AHN (ex 16. primi) foret acutus, propter suppositum rectum angulum externum AND; ac propterea angulus DHL foret obtusus, et sic in quadrilatero DHLK quatuor haberemus angulos, qui simul sumpti majores forent quatuor rectis, contra praedictam hypothesin anguli acuti. Igitur constat ab ea AL secari debere angulum BAN, qui propterea nequit dici omnium minimus, sub quo educta AN commune habeat cum illa BX in duobus distinctis punctis perpendiculum ND. Quod erat secundo loco demonstrandum Itaque constat etc.

COROLLARIUM.

Inde autem observare licet, quod sub angulo minore BAL obtinetur (in hypothesi anguli acuti) commune LK perpendiculum, remotius quidem ab illa basi AB, prout constat ex ipsa constructione, sed rursum minus altero viciniore communi perpendiculo ND, quod obtinetur sub angulo majore BAN. Ratio hujus posterioris est, [67] quia in quadrilatero LKDS angulus ad punctum S acutus est in praedicta hypothesi, cum reliqui tres supponantur recti. Quare (ex Cor. I. post 3. hujus) latus LK minus erit contraposito latere SD, et sic multo minus latere ND.

But again AL cannot be the continuation of AN; because otherwise in the quadrilateral NDKL we would have four right angles, against the hypothesis of acute angle.

But neither can it cut DN produced in any exterior point H; because angle AHN (from Eu. I. 16) would be acute, on account of the external angle AND supposed right; and therefore angle DHL would be obtuse, and so in the quadrilateral DHLK we would have four angles, which taken together would be greater than four right angles, against the aforesaid hypothesis of acute angle.

Therefore it follows that the angle BAN must be cut by this AL, and therefore cannot be declared the least of all, drawn under which AN has with BX in two distinct points a common perpendicular ND.

Quod erat secundo loco demonstrandum. Itaque constat etc.

COROLLARY.

But hence is permitted to observe, that under a lesser angle BAL is obtained (in the hypothesis of acute angle) a common perpendicular LK, more remote indeed from the base AB, as follows from the construction, but moreover less than the other nearer common perpendicular ND, which is obtained under a greater angle BAN.

The reason of this latter is [67] because in the quadrilateral LKDS the angle at the point S is acute in the aforesaid hypothesis, since the three remaining angles are supposed right.

Wherefore (from Cor. I. to P. III.) the side LK will be less than the opposite side SD, and so much less than the side ND.

PROPOSITIO XXXI.

Jam dico nullum fore praedictorum in duobus distinctis punctis communium perpendiculorum limitem determinatum, quo minus sub minore, ac minore acuto angulo, ad illud punctum A constituto, deveniri semper possit (juxta hypothesin anguli acuti) ad tale commune in duobus distinctis punctis perpendiculum, quod sit minus qualibet assignata longitudine R.

Demonstratur. Quatenus enim aliter res se habeat; si ex puncto K (recole fig. 30.) in quantalibet a puncto B distantia in ea BX assignato, educatur perpendicularis KL, ad quam ex puncto A (juxta 12. primi) demissa intelligatur perpendicularis AL, deberet ipsa KL major esse ea longitudine R. Ratio autem est; quia assumpto in eadem BX altiore puncto Q, ex quo educatur ad ipsam BX perpendicularis QF, ad quam (juxta eandem 12. primi) demittatur perpendicularis AF, deberet haec rursum saltem non esse minor ea longitudine R. Erit autem KL (ex Cor. praeced. Prop.) major ipsa QF. Igitur ea KL major foret praedicta longitudine R. Atque ita semper altius procedendo.

Jam vero: si illa quantacunque KB divisa intelligatur (prout in XXV. hujus) in portiones KK, aequales illi longitudini R, educanturque ex illis punctis K perpendiculares, quae occurrant ipsi AX in punctis H, D, M; non erunt anguli ad haec puncta, versus partes puncti L, aut recti, aut obtusi; ne in aliquo quadrilatero, ut puta

PROPOSITION XXXI.

Now I say there will be, of the aforesaid common perpendiculars in two distinct points, no determinate limit, such that under a smaller and smaller acute angle made at the point A, it would not always be possible to attain (in the hypothesis of acute angle) to such a common perpendicular in two distinct points as is less than any assignable length R.

PROOF. For in so far as the thing were otherwise; if from the point K (resume fig. 30) in BX assigned at any however great distance from the point B, a perpendicular KL is erected, to which from point A (by Eu. I. 12) the perpendicular AL is supposed let fall, KL ought to be greater than the length R.

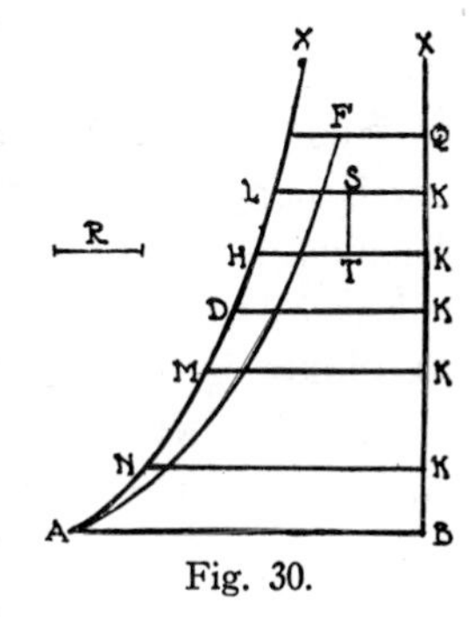

Fig. 30.

The reason is; because a higher point Q being assumed in this BX, from which is erected to BX the perpendicular QF, to which (by the same Eu. I. 12) a perpendicular AF is let fall, this again must anyhow not be less than the length R.

But KL (from Cor. to preceding proposition) will be greater than QF. Therefore KL would be greater than the aforesaid length R. And so ever proceeding higher.

But now, if this however great KB is supposed divided (as in P. XXV.) into portions KK, equal to the length R, and from these points K perpendiculars are erected, which meet AX in points H, D, M; the angles at these points, toward the parts of the point L, will neither be right nor obtuse; lest in some quadrilateral, as suppose

KM-[68]LK quatuor simul anguli aequales sint, aut majores quatuor rectis, contra hypothesim anguli acuti, juxta quam procedimus. Omnes igitur hujusmodi anguli acuti erunt versus partes puncti L; ac propterea omnes itidem ad illa puncta obtusi versus partes puncti A. Quare (ex Cor. I. post 3. hujus) praedictarum perpendicularium minima quidem erit KL remotior a basi AB, maxima KM propinquior eidem basi; reliquarum vero propinquior remotiore semper major erit. Igitur (ex mea praeced. 24. ejusque Coroll.) quatuor simul anguli quadrilateri KHLK remotioris a basi AB majores erunt quatuor simul angulis reliquorum omnium quadrilaterorum eidem basi proximiorum. Quare (prout XXV. hujus) destructa maneret hypothesis anguli acuti.

Itaq; constat nullum fore praedictorum in duobus distinctis punctis communium perpendiculorum limitem dederminatum, quo minus sub minore, ac minore acuto angulo, ad illud punctum A constituto, deveniri semper possit (juxta hypothesin anguli acuti) ad tale commune in duobus distinctis punctis perpendiculum, quod sit minus qualibet assignata longitudine R. Quod erat etc.

PROPOSITIO XXXII.

Jam dico unum aliquem fore (in hypothesi anguli acuti) determinatum acutum angulum BAX, sub quo educta AX (fig. 33.) non nisi ad infinitam distantiam incidat in eam BX, ac propterea sit ipsa limes partim intrinsecus, partim extrinsecus; tum earum omnium, quae sub minoribus acutis angulis ad finitam distantiam incidunt in praedictam BX; tum etiam aliarum,

KMLK, [68] the four angles together should be equal to or greater than four rights, contrary to the hypothesis of acute angle, according to which we are proceeding. Therefore all such angles will be acute toward the parts of the point L; and therefore in like manner all at these points obtuse toward the parts of the point A. Wherefore (from Cor. I to P. III.) of the aforesaid perpendiculars the least will indeed be KL more remote from the base AB, the greatest KM nearer this base.

And of the remaining the nearer will be ever greater than the more remote.

Therefore (from the preceding P. XXV, and its corollary) the four angles together of the quadrilateral KHLK more remote from base AB will be greater than the four angles together of all the remaining quadrilaterals nearer to this base. Wherefore (as in P. XXV.) the hypothesis of acute angle would be destroyed.

Therefore it holds, that of the aforesaid common perpendiculars in two distinct points there will be no determinate limit, such that under a smaller and smaller acute angle made at the point A, it would not always be possible to attain (in the hypothesis of acute angle) to such a common perpendicular in two distinct points as may be less than any assigned length R.

Quod erat demonstrandum.

PROPOSITION XXXII.

Now I say there is (in the hypothesis of acute angle) a certain determinate acute angle BAX drawn under which AX (fig. 33) only at an infinite distance meets BX, and thus is a limit in part from within, in part from without; on the one hand of all those which under lesser acute angles meet the aforesaid BX at a finite distance; on the other hand also of the others

quae sub majoribus angulis acutis, usque ad angulum rectum inclusive, commune obtinent in duobus distinctis punctis perpendiculum cum eadem BX. [69]

Demonstratur. Nam primo constat (ex Cor. II. post XXIX. hujus) nullum fore determinatum acutum angulum, omnium maximum, sub quo educta ex illo puncto A ad finitam distantiam occurrat praedictae BX. Secundo constat nullum itidem esse (in hypothesi anguli acuti) acutum angulum omnium minimum, sub quo educta commune habeat in duobus distinctis punctis perpendiculum cum illa BX; quandoquidem (ex praecedente) nullus esse potest limes determinatus, quo minus sub minore acuto angulo ad illud punctum A constituto deveniri possit ad tale commune in duobus distinctis punctis perpendiculum, quod sit minus qualibet assignabili longitudine R.

Atque hinc tertio consequitur unum aliquem (in ea hypothesi) esse debere determinatum acutum angulum BAX, sub quo educta AX ita semper magis accedat ad eam BX, ut non nisi ad infinitam distantiam in eandem incidat.

Porro autem hanc ipsam AX fore limitem partim intrinsecum, partim extrinsecum utriusque praedictarum rectarum classis, sic demonstratur. Nam primo conveniet cum illis rectis, quae ad finitam distantiam occurrunt ipsi BX, cum ipsa etiam aliquando conveniat; discrepabit autem, quia ipsa non nisi ad infinitam distantiam. Secundo autem conveniet etiam, et simul discrepabit ab illis rectis, quae commune obtinent in duobus distinctis punctis perpendiculum cum illa BX; quia ipsa etiam commune obtinet perpendiculum cum eadem BX; sed in uno eodemque puncto X infinite dissito. Hoc autem postremum censeri

which under greater acute angles, even to a right angle inclusive, have a common perpendicular in two distinct points with BX. [69]

PROOF. First it holds (from Cor. II. to P. XXIX.) that no determinate acute angle will be the greatest of all drawn under which a straight from the point A meets the aforesaid BX at a finite distance.

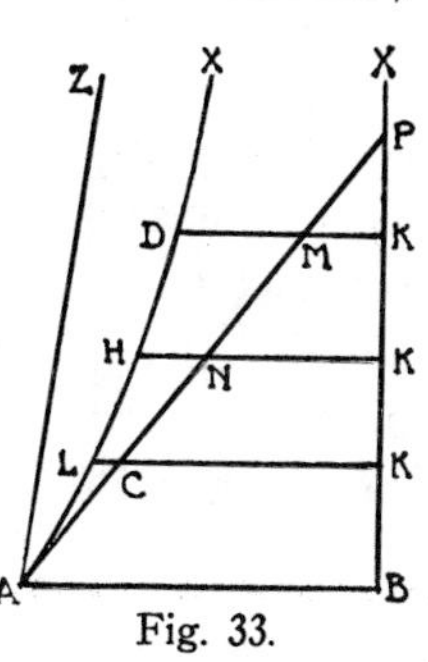

Fig. 33.

Secondly, it holds in like manner that (in the hypothesis of acute angle) no acute angle will be the least of all drawn under which a straight has a common perpendicular in two distinct points with BX; since indeed (from what precedes) there can be no determinate limit, such that there cannot be found, under a lesser angle constituted at the point A, a common perpendicular in two distinct points, which is less than any assignable length R.

And hence follows thirdly, that (in this hypothesis) there must be a certain determinate acute angle BAX, drawn under which AX so approaches ever more to BX, that only at an infinite distance does it meet it.

But further that this AX is a limit in part from within in part from without of each of the aforesaid classes of straights is proved thus. First, it agrees with those straights which meet BX at a finite distance since it also finally meets; but it differs, because it meets only at an infinite distance.

But secondly it also agrees with, and at the same time differs from those straights which have a common perpendicular in two distinct points with BX; because it also has a common perpendicular with BX; but in one and the same point X infinitely distant. But this latter ought to

debet demonstratum in XXVIII. hujus, prout moneo in ejusdem Corollario.

Itaque constat unum aliquem fore (in hypothesi anguli acuti) determinatum acutum angulum BAX, sub quo educta AX non nisi ad infinitam distantiam incidat in [70] eam BX, ac propterea sit ipsa limes partim intrinsecus, partim extrinsecus; tum earum omnium, quae sub minoribus acutis angulis ad finitam distantiam incidunt in praedictam BX; tum etiam aliarum, quae sub majoribus angulis acutis, usque ad angulum rectum inclusive, commune obtinent in duobus distinctis punctis perpendiculum cum eadem BX. Quod erat etc.

PROPOSITIO XXXIII.

Hypothesis anguli acuti est absolute falsa; quia repugnans naturae lineae rectae.

Demonstratur. Ex praemissis Theorematis constare potest eo tandem perducere Geometriae Euclideae inimicam hypothesin anguli acuti, ut agnoscere debeamus duas in eodem plano existentes rectas AX, BX, quae in infinitum protractae versus eas partes punctorum X in unam tandem eandemque rectam lineam coire debeant, nimirum recipiendo, in uno eodemque infinite dissito puncto X, commune in eodem cum ipsis plano perpendiculum. Quoniam vero de primis ipsis principiis agendum mihi hic est, diligenter curabo, ut nihil omittam quasi nimis scrupulose objectum, quod quidem exactissimae demonstrationi opportunum esse cognoscam.

LEMMA I.

Duae rectae lineae spatium non comprehendunt.

Definit Euclides lineam rectam, quae *ex aequo sua interjacet puncta.* Esto igitur (fig. 37.) linea quaedam

be considered demonstrated in P. XXVIII., as I point out in its corollary.

Therefore it holds, that (in the hypothesis of acute angle) there will be a certain determinate acute angle BAX, drawn under which AX only at an infinite distance meets [70] BX, and thus is a limit in part from within, in part from without; on the one hand of all those which under lesser acute angles meet the aforesaid BX at a finite distance; on the other hand also of the others which under greater acute angles, even to a right angle inclusive, have a common perpendicular in two distinct points with BX.

Quod erat etc.

PROPOSITION XXXIII.

The hypothesis of acute angle is absolutely false; because repugnant to the nature of the straight line.

Proof. From the foregoing theorem may be established, that at length the hypothesis of acute angle inimical to the Euclidean geometry has as outcome that we must recognize two straights AX, BX, existing in the same plane, which produced *in infinitum* toward the parts of the points X must run together at length into one and the same straight line, truly receiving, at one and the same infinitely distant point a common perpendicular in the same plane with them. ✡

But since I am here to go into the very first principles, I shall diligently take care, that I omit nothing objected almost too scrupulously, which indeed I recognize to be opportune to the most exact demonstration. ✡

LEMMA I.

Two straight lines do not inclose a space.

Euclid defines a straight line as one which *lies evenly between its points.*

AX, quae ex puncto A per sua quaelibet intermedia puncta continuative excurrat usque ad punctum X. Non di-[71]cetur haec linea recta, si talis ipsa fuerit, ut circa duo illa immota extrema sua puncta possit ipsa in alteram partem converti, ut puta a laeva parte in dexteram: Non dicetur, inquam, linea recta; quia non jacebit ex aequo inter sua designata extrema puncta; quandoquidem vel in laevam partem declinabit, ubi ex puncto A excurrit ad punctum X per quaedam intermedia puncta B; vel declinabit in dexteram, ubi ex eodem immoto puncto A excurrit ad idem immotum punctum X per quaedam intermedia puncta C, quae alia plane sunt a praedictis punctis B. Scilicet illa sola linea AX dici poterit recta, quae excurrat ex puncto A ad punctum X per talia intermedia puncta D, quae ipsa, prout sic invicem continuata, revolvi nequeant, circa illa immota extrema puncta A, et X, ad novum et novum occupandum situm.

In hac autem rectae lineae idea manifeste continetur proposita veritas, duas nempe rectas lineas spatium non comprehendere. Si enim duae exhibeantur lineae claudentes spatium, quarum nempe communia sint extrema duo puncta A, et X, facile ostenditur vel neutram, vel unam tantum illarum linearum esse rectam. Neutra erit recta, ut puta ABBX, et ACCX, si circa duo extrema immota puncta A, et X, ita revolvi posse intelligantur ipsae ABBX, ACCX, ut reliqua ipsarum intermedia puncta ad novum, et novum occupandum locum pertranseant. Una tantum erit recta, ut puta ADDX, si circa illa immota extrema puncta ita revolvi intelligantur ipsae ABBX, ACCX, quae hinc inde cum illa ADDX spatium claudunt, ut ipsarum quidem ABBX, ACCX puncta intermedia ad

Let there be therefore (fig. 37) any line AX, which from the point A through any intermediate points of it runs consecutively even to the point X. This line is not [71] called straight if it be such, that it can be turned about its two end points into another region, as suppose from the left region into the right: I say it is not called a straight line; because it will not lie *ex aequo* between its designated extreme points; since either it will lean toward the left side, where from the point A it runs out to the point X through certain intermediate points B; or it bends to the right, where from the same fixed point A it runs out to the same fixed point X through certain intermediate points C which are wholly different from the aforesaid points B. Obviously only that line AX can be called straight, which runs out from the point A to the point X through such intermediate points D, as, in order one after another continued, cannot be revolved, about those fixed extreme points A, and X, to occupying new and new location.

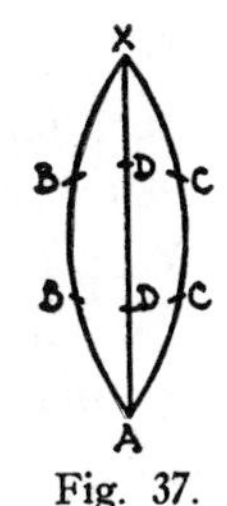

Fig. 37.

But in this idea of the straight line is contained manifestly the announced truth, namely that two straight lines do not inclose a space. For if two lines are shown inclosing a space, which have in common the two extreme points A, and X, it is easily shown either that neither, or only one of them is straight. Neither will be straight, as for example ABBX, and ACCX, if it be supposed so that they can be revolved about two fixed extreme points A, and X, that their remaining intermediate points pass over to occupying new and new place.

One only will be straight, as for example ADDX, if about those fixed end points we may suppose ABBX, ACCX, which on both sides with that ADDX inclose a space, so to be revolved, that indeed the intermediate points

novum, et novum occupandum locum pertranseant, ipsius vero ADDX puncta omnia etiam intermedia in eodem loco persistant. Non igitur fieri potest, ut duae juxta praemissam intelligentiam rectae lineae, spatium comprehendant. Quod erat propositum. [72]

COROLLARIUM I.

Hinc porro sequitur admitti oportere postulatum illud Euclidaeum: quod *a dato puncto ad quodlibet assignatum punctum rectam lineam ducere liceat.* Nam clare intelligitur, duas semper sine ullo certo limite duci posse lineas, praedictis punctis A, et X terminatas, quae propiores invicem fiant, minusque idcirco spatium comprehendant, dum scilicet una quidem ducatur ad laevam partem, et altera uniformis ad dexteram, sive una sursum, et altera deorsum; duci, inquam, posse lineas ejusmodi semper invicem sine ullo certo limite propiores, quae utique omnino uniformes inter se sint, sibique invicem idcirco succedant, dum circa immota extrema puncta A, et X, revolvi ipsae intelligantur. Inde autem clare itidem intelligitur, sequi tandem debere (in semper majore harum uniformium linearum, unius ad alteram accessu) coitionem in unam, eandemque lineam ADX, quae circa immota extrema illa puncta revolvi nequeat ad occupandum novum locum. Et haec erit linea recta postulata.

Ubi rursum constat unicam esse, quae a dato puncto ad quodlibet alterum assignatum punctum potest duci linea recta.

of ABBX, ACCX, pass over to the occupying of new and new position, but on the contrary all the intermediate points of ADDX remain in the same place.

Therefore it cannot be, that two lines, straight in accordance with the previous concept, inclose a space.

Quod erat propositum. [72]

COROLLARY I.

Hence moreover follows we should admit the Euclidean postulate: that *from a given point to any assigned point a straight line may be drawn.* ✡

For it is clearly understood, that always two lines without any certain limit can be drawn, terminated in the aforesaid points A, and X, which mutually approach, and therefore inclose less space, while indeed one is drawn toward the left side, and the other of the same shape toward the right, or one over, and the other under; I say, lines of this sort may be drawn always mutually approaching without any certain limit, which are completely of the same shape with each other, and therefore mutually succeed each other when supposed revolved about the fixed end points A, and X.

Whence clearly in like manner is understood, at length (in ever greater approach of these like shaped lines, one to the other) should follow the coalescence into one, and the same line ADX, which cannot be revolved about those fixed extreme points so as to occupy a new position. And this will be the straight line postulated.

Where again is established to be unique the straight line, which can be drawn from a given point to any other assigned point.

COROLLARIUM II.

Praeterea sequitur uniformem esse debere intelligentiam alterius Euclideae definitionis, in qua dicit planam superficiem esse, *quae ex aequo suas interjacet lineas.* Si enim superficies clausa praedictis lineis una ADX recta, et altera ABBX (sive haec sit unica, aut multiplex linea curva, sive sit composita ex duabus, aut pluribus lineis rectis, ut puta AB, BB, BX) si, inquam, superficies [73] ejusmodi revolvi intelligatur circa immotam rectam ADX, usque dum ipsa linea ABX perveniat ad congruendum lineae ACX, in parte adversa locatae, quae utique ad omnimodam aequalitatem, et similis omnino sit ipsi ABX, et rursum cum eadem recta ADX claudat (versus eandem sive supernam, sive infernam partem) superficiem omnino aequalem, et similem antedictae; alterutrum sane continget; vel ita ut una superficies alteri adamussim congruat; vel ita ut intra duas illas superficies claudatur spatium trinae dimensionis. Et primum quidem si contingat, dicetur superficies plana; sin vero contingat secundum, non dicetur superficies plana; quia tunc aliae intermediae intelligi poterunt inter easdem extremas lineas interpositae superficies invicem aequales, ac similes, quae semper magis ad se invicem sine ullo certo limite accedant, ac propterea usque ad excludendum omne spatium intermedium. Tunc autem utraque illa superficies dicetur plana, quia vere jacebit ex aequo inter suas extremas lineas, sine ullo ascensu, aut descensu in partes adversas.

LEMMA II.

Duae lineae rectae non possunt habere unum et idem segmentum commune.

Demonstratur. Si enim fieri potest; unum et idem segmentum AX commune sit (fig. 38.) duabus rectis,

COROLLARY II.

Moreover it follows the interpretation should be the same of the other Euclidean definition, in which he says a surface is plane, *which lies evenly between its lines.*

For if a surface inclosed by the aforesaid lines one ADX straight, and another ABBX (whether this be a simple or multiplex curved line, or be composed of two, or several straight lines, as suppose AB, BB, BX) if, I say, a surface [73] of this sort is supposed to be revolved about the fixed straight ADX, until the line ABX comes to congruence with the line ACX, located in the opposite part, which assuredly is in every way equal and wholly similar to ABX, and again with the same straight ADX incloses (toward the same part, whether upper or under) a surface wholly equal, and similar to the aforesaid: one of two things certainly happens; either one surface fits the other completely; or between those two surfaces is inclosed a three-dimensional space.

And indeed if the first happens, the surface is called plane; but if the second happens the surface is not called plane; because then may be supposed other intermediate surfaces, mutually equal, and similar, interposed between the same extreme lines, which always mutually approach more to each other without any certain limit, and therefore even to the exclusion of every intermediate space.

But then each surface is called plane, because truly it lies *ex aequo* between its extreme lines, without any ascent or descent into bordering parts.

LEMMA II.

Two straight lines cannot have one and the same segment in common.

PROOF. For if that is possible, let one and the same segment AX be common (fig. 38) to the two straights

per punctum X in eodem plano continuatis AXB, et AXC. Tum centro X, et intervallo XB, sive XC, describatur arcus BMC, ad cujus quodlibet punctum M jungatur ex puncto X recta XM.

Dico primo, lineam AXM fore et ipsam, in facta hy-[74]pothesi, lineam rectam, ex puncto A per punctum X continuatam. Si enim linea ejusmodi recta non sit, duci poterit (ex Cor. I. praecedentis Lemmatis) alia quaedam linea AM, quae ipsa sit recta. Haec autem vel secabit in aliquo puncto K alterutram ipsarum XB, XC; vel earundem alterutram, ut puta eam XB claudet intra spatium comprehensum ipsis AX, XM, et APLM. At horum prius manifeste repugnat praecedenti Lemmati; quia sic duae suppositae rectae lineae, una AXK, et altera ATK, spatium clauderent. Posterius autem uniformis absurdi statim convincitur.

Nam constat rectam XB, si per B ulterius protrahatur, occursuram tandem in aliquo puncto L ipsi APLM; unde rursum duae suppositae rectae, una AXBL, et altera APL, spatium claudent. Porro uniforme sequitur absurdum, si fingamus, quod recta XB, ulterius protracta per B, occurrat tandem in quovis alio puncto aut rectae XM, aut rectae XA.

Ex istis autem evidenter consequitur lineam AXM fore ipsam, in facta hypothesi, lineam rectam ex puncto A ad punctum M deductam. Quod erat propositum.

Dico secundo, eam suppositam rectam AXB (quatenus quidem intelligatur conservare suam illam qualem-

AXB, and AXC produced through the point X in the same plane. Then with center X, and radius XB, or XC, describe the arc BMC, to any point of which M is drawn from the first point X the straight XM.

I say first, under the assumed hypothesis also the line AXM will be [74] a straight line, continued from the point A through the point X.

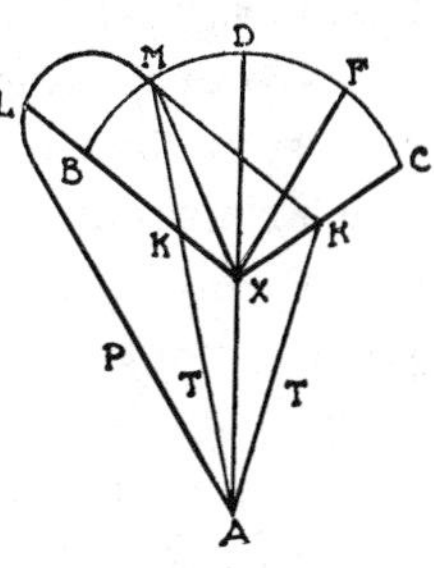

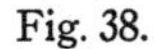
Fig. 38.

For if a line of this sort be not straight, there can be drawn (from Cor. I. of the preceding lemma) a certain other line AM, which itself is straight. But this either cuts in some point K one or the other of those straights XB, XC; or it incloses one or the other of them, as suppose XB within the space bounded by AX, XM, and APLM.

But the first of these is manifestly contrary to the preceding lemma; because thus two lines supposed straight, one AXK, and the other ATK, would inclose a space.

But the second is at once convicted of a like absurdity.

For it is certain that the straight XB, if produced on through B, will at length meet this APLM in some point L; whence again two lines supposed straight, one AXBL, and the other APL, will inclose a space. But a like absurdity follows, if we assume, that the straight XB, produced on through B, at length meets in some other point either the straight XM, or the straight XA.

But from this evidently follows that the line AXM is itself, in the assumed hypothesis, the straight line drawn from the point A to the point M.

Quod erat propositum.

I say secondly, that the assumed straight AXB (inasmuch as it is understood to retain its arbitrary continuation

cunque continuationem ex puncto A per X versus B) non posse recipere duplicem aliam in eodem plano positionem, in quarum utraque portio quidem AX in eodem situ persistat, portio vero altera XB in una illarum duarum positionum congruat (exempli causa) ipsi XC, et in alia positione congruat ipsi XM.

Scilicet non hic renuo, quin portio XB, si intelligatur moveri in illo suo plano circa punctum X, adeo ut successive adamussim congruat (ex praecedente Lemmate) non modo ipsis XM, XC, verum etiam adamussim con-[75] gruat infinitis aliis rectis, quae ex puncto X duci possunt ad reliqua intermedia puncta arcus BC: Non, inquam, hic renuo, quin illa XB in qualibet illarum positionum considerari debeat tanquam continuatio in rectum ipsius immotae AX; cum magis circa eam AXM jam demonstraverim id secuturum in facta hypothesi illius communis segmenti: Unice igitur hic assero, in una tantum novarum illarum positionum, ut puta dum congruit ipsi XC, retineri ab ea posse illam eandem qualemcunque continuationem, quam obtinet in prima positione, ubi ex puncto A per X procedit versus punctum B.

Et istud quidem sic demonstratur. Nam primo constat continuationem illam AXB nequire esse omnino similem, aut aequalem continuationi AXC, si utraque consideretur versus eandem seu laevam, seu dexteram partem; quia caeterum in ea tali positione deberent invicem congruere ipsae AXB, AXC; quod est contra hypothesim communis illius segmenti AX: Deberent, inquam, congruere; dum scilicet, relate ad eam immotam AX, aeque similiter in eandem seu laevam, seu dexteram partem convergerent in eo tali plano illae continuatae XB, et XC. Secundo constat nihil vetare, quin praedicta continuatio AXB, considerata versus unam partem, ut puta, ad laevam, similis plane sit, aut aequalis continuationi AXC, consideratae versus partem adversam, ut puta, ad dexte-

from the point A through X toward B) cannot have two different positions in the same plane, in both of which the portion indeed AX persists in the same place, but the other portion XB in one of those two positions fits (for example) XC, and in the other position fits XM. ✡

Of course I do not here deny, that the portion XB, if it is supposed to be moved in its plane about the point X, so that successively it fits exactly (from the preceding lemma) not merely XM, XC, but also exactly fits [75] the other infinitely many straights, which from the point X may be drawn to the remaining intermediate points of the arc BC: I say, I do not here deny, that XB in any of its positions may be considered as the continuation in a straight of that fixed AX; when rather I have demonstrated already about AXM that this would happen in case of the hypothesis of a common segment: Solely therefore I here affirm, in one merely of those new positions,✡ as suppose while it fits XC, may be retained by it the same arbitrary continuation, which it has in the first position, where from the point A it goes out through X toward the point B.

And this indeed is demonstrated thus. For first it is evident that the continuation AXB cannot be wholly similar, or equal to the continuation AXC, if each is considered toward the same part whether left or right; because otherwise in such position AXB, AXC must mutually coincide; which is against the hypothesis of that common segment AX: I say, must coincide; provided that of course, in relation to the same fixed AX, the continuations XB, and XC in the plane concerned extend just similarly toward the same part whether left or right.

Secondly is evident that nothing prevents the aforesaid continuation AXB, considered toward one part, as suppose, toward the left, being precisely similar, or equal to the continuation AXC, considered toward the opposite part,

ram, adeo ut propterea, sine ulla immutatione in ipsa AXB, locari haec possit ad congruendum in eodem plano alteri AXC. At manifeste repugnat, quod rursum, sine ulla immutatione illius suae continuationis, locari ea possit in eodem plano ad congruendum alteri AXM, quae nimirum dividat in X illum qualemcunque angulum BXC. Quod enim continuatio AXB alia plane sit a continuatione AXM, si utraque consideretur versus eandem seu lae-[76] vam, seu dexteram partem, ex eo manifestum esse debet; quia caeterum (ut in simili observatum jam est) in ea tali positione deberent invicem congruere ipsae AXB, AXM. Sed neque sustineri potest, quod continuatio AXB versus unam partem, ut puta ad laevam, similis plane sit, aut aequalis continuationi AXM versus partem adversam, ut puta ad dexteram; quia caeterum continuatio AXM versus dexteram similis plane foret, aut aequalis continuationi AXC versus eandem dexteram partem propter suppositam omnimodam similitudinem, aut aequalitatem inter modo dictam continuationem, et illam aliam AXB versus laevam. Tunc autem in ea tali positione (ut est praedictum) deberent invicem congruere ipsae AXM, AXC; quod est contra praesentem hypothesim.

Ex quibus omnibus infero: eam suppositam rectam AXB (quatenus quidem intelligatur conservare suam illam qualemcunque continuationem ex puncto A versus B) recipere non posse duplicem aliam in eodem plano positionem, in quarum utraque portio quidem AX in eodem situ persistat, portio vero altera XB in una illarum duarum positionum congruat (exempli causa) ipsi XC, et in alia positione congruat ipsi XM. Quod erat propositum.

Dico tertio: eandem suppositam rectam AXB non

as suppose, toward the right, so that consequently, without any change in AXB, this may be brought to congruence with the other AXC in the same plane.

But it is manifestly contradictory, that on the other hand, without any change of its prolongation, this can be brought in the same plane into congruence with the other AXM, which indeed at X divides that arbitrary angle BXC.

For that the prolongation AXB is plainly other than the prolongation AXM, if each is considered toward the same part, whether left [76] or right, must be manifest from this; because otherwise (as already observed in like case) in such a situation AXB, AXM must mutually fit.

But neither can it be maintained, that the prolongation AXB toward one part, as suppose toward the left, is wholly similar, or equal to the prolongation AXM toward the opposite part, as suppose toward the right; because otherwise the prolongation AXM toward the right would plainly be similar, or equal to the prolongation AXC toward the same right side, because of the assumed complete similitude, or equality between the just cited prolongation, and that other AXB toward the left.

But then in such a situation (as previously remarked) AXM, AXC should mutually fit; which is against the present hypothesis.

From all which, I infer: the assumed straight AXB (in so far as it is understood to retain its arbitrary prolongation from the point A toward B) cannot have two different positions in the same plane, in both of which the portion indeed AX remains in the same location, but the other portion XB in one of those two positions fits (for example) XC, and in the other position fits XM.

Quod erat propositum.

I say thirdly: the assumed straight AXB can in no

alia ratione conservare posse suam illam qualemcunque continuationem, dum ejusdem portio XB intelligitur transferri per nova, et nova loca usque ad congruendum in illo quodam plano ipsi XC, persistente interim in eodem suo loco portione AX; non posse, inquam, conservare suam illam qualemcunque continuationem, nisi quatenus portio ipsa XB intelligatur ascendere, aut descendere ad existendum cum illa immota AX in novis, et novis planis, usque dum redeat ad antiquum planum, congruens ibi praedictae XC.[77]

Id enim censeri potest jam demonstratum; quia scilicet nulla alia in eodem illo plano reperiri potest positio, juxta quam ipsa AXB (persistente portione AX in suo eodem loco) conservet suam illam qualemcunque continuationem, praeterquam ubi deveniat ad congruendum praedictae AXC.

Dico quarto: designari posse in eo arcu BC tale punctum D, ad quod si jungatur XD, jam ipsa AXD non modo recta linea sit, sed rursum ita se habeat, ut continuatio AXD, considerata versus laevam, aequalis plane sit, aut similis eidem continuationi consideratae versus dexteram.

Demonstratur. Et prior quidem pars (qualecunque sit illud punctum D in arcu BC designatum) eo modo ostenditur, quo supra usi sumus circa continuatam AXM. Posterior vero pars ita evincitur. Nam hic supponimus duas rectas AXB, AXC, sub eodem communi segmento AX. Praeterea supponimus continuationem AXB versus laevam non esse omnino similem, aut aequalem eidemmet continuationi versus dexteram; quia stante omnimoda ejusmodi similitudine, aut aequalitate, facile ostenditur nulli alteri rectae lineae commune esse posse illud segmentum AX, prout nempe sic demonstrabimus de illa continuata AXD. Tandem consequenter supponimus continuatam illam AXB ita locari posse in eodem plano, ut

other way retain its arbitrary prolongation, while its part XB is supposed to be transferred through new and new positions even to fitting XC in that one plane, the portion AX remaining meanwhile in the same place; I say it cannot retain its chosen continuation, except in so far as the portion XB is understood to ascend, or to descend to be with the fixed AX in new, and new planes, until it returns to the old plane, fitting there the aforesaid XC. [77]

For this may be adjudged already demonstrated; because obviously no position in that same plane can be found, at which AXB (the portion AX remaining in its place) retains its chosen prolongation, except where it comes to congruence with the aforesaid AXC.

I say fourthly: in the arc BC such a point D can be designated that, if XD be joined, then this AXD not only is a straight line, but moreover it lies so, that the prolongation AXD, considered toward the left, is wholly equal, or similar to the same prolongation considered toward the right.

Proof. The first part (whatever be the point D designated in the arc BC) is shown by the method used above in regard to the prolongation AXM.

But the second part is proved thus. We suppose here two straights AXB, AXC with the same common segment AX. Further we suppose the prolongation AXB toward the left not to be wholly similar, or equal to the same prolongation toward the right; because, such a complete similitude or equality holding good, it is easily shown that segment AX can be common to no other straight line, just as we shall demonstrate of the prolongation AXD. Finally in consequence we suppose the prolongation AXB may so

sub eodem immoto segmento AX congruat cuidam alteri AXC, in qua nimirum continuatio ipsa AXC versus dexteram similis plane sit, aut aequalis continuationi AXB versus laevam, ac rursum continuatio AXC versus laevam similis plane sit, aut aequalis continuationi AXB versus dexteram.

His stantibus: si ad quodvis punctum M sumptum in eo arcu BC jungatur XM; vel continuatio AXM erit [78] sibi ipsi plane uniformis relate ad laevam, ac dexteram partem ipsius AX; vel non. Si primum; demonstrabo de ista AXM, quod statim demonstraturus sum de illa continuata AXD. Si secundum, ergo praedicta AXM ita rursum locari poterit in eodem plano, ut sub eodem immoto segmento AX congruat cuidam alteri AXF, in qua nimirum continuatio ipsa AXF versus dexteram similis plane sit, aut aequalis continuationi AXM versus laevam, ac rursum continuatio AXF versus laevam similis plane sit, aut aequalis continuationi AXM versus dexteram. Porro, cum punctum M supponi possit vicinius puncto B, quam punctum C, non cadet punctum F in ipsum punctum C; quia sic continuatio AXM versus laevam similis plane foret, aut aequalis continuationi AXF, sive AXC versus dexteram, ac propterea similis plane, aut aequalis continuationi AXB versus laevam, quod est absurdum, cum illae duae XM, XB, non sibi invicem congruant in sua tali positione. Sed neque etiam existet punctum F ultra punctum C in eo arcu BC ulterius producto; quia sic uniformi ratiocinio ostendetur, contra hypothesim, quod etiam punctum M deberet existere in eo arcu CB ulterius producto, adeo ut nimirum ipsa XM divideret versus laevam eum qualemcunque angulum AXB, prout XF poneretur dividere versus dexteram eum qualemcunque angulum AXC: Deberet, inquam, sic existere, ad eum utique

be located in that plane, that with its fixed segment AX it fits a certain other, AXC, in so far as truly the prolongation AXC toward the right is exactly similar, or equal to the prolongation AXB toward the left, and moreover the prolongation AXC toward the left is precisely similar, or equal to the prolongation AXB toward the right.

This remaining: if, assuming any point M in the arc BC, we join XM; either the prolongation AXM will be [78] precisely uniform in relation to the left, and the right side of AX; or not. If the first; I shall demonstrate of AXM, what immediately I shall have demonstrated of the prolongation AXD. If the second, therefore the aforesaid AXM can in turn be so located in the same plane, that with the same fixed segment AX it fits a certain other AXF, in which truly the prolongation AXF toward the right is precisely similar, or equal to the prolongation AXM toward the left, and moreover the continuation AXF toward the left is precisely similar, or equal to the prolongation AXM toward the right.

Furthermore, since the point M may be supposed nearer to the point B than is the point C, the point F does not fall upon the point C; because thus the prolongation AXM toward the left would be precisely similar, or equal to the prolongation AXF, or AXC toward the right, and therefore precisely similar, or equal to the prolongation AXB toward the left, which is absurd, since the two XM, XB do not mutually fit each other in such position of theirs.

But neither also is the point F beyond the point C in the arc BC produced farther on; because thus by like reasoning is shown, against the hypothesis, that also the point M must be in the arc CB produced farther on, so that XM would divide toward the left the arbitrary angle AXB, just as XF would be posited to divide toward the right the arbitrary angle AXC: I say must so lie, to the

finem, ut ea AXM sub eodem immoto segmento AX locari rursum possit in eodem plano ad congruendum illi alteri AXF, in qua nimirum continuatio ipsa AXF versus dexteram similis plane sit, aut aequalis continuationi AXM versus laevam, ac rursum continuatio AXF versus laevam similis plane sit, aut aequalis continuationi AXM versus dexteram.

Quoniam vero arcus BC major est ejusdem portione [79] MF, designarique uniformiter possunt in ea portione MF alia duo puncta cum minore, sine ullo certo termino, intercapedine; alterutrum sane in hac praedictorum punctorum approximatione contingere debet. Unum est, si tandem incidatur in unum idemque intermedium punctum D, ad quod si jungatur XD, talis habeatur continuatio AXD, cui soli conveniat (facta comparatione inter laevam, ac dexteram partem) esse sibi ipsi omnino similem, aut aequalem. Alterum est, si duo talia inveniantur distincta puncta M, et F, ad quae junctae XM, et XF, duas exhibeant continuationes, unam AXM, et alteram AXF, quarum utraque sit sibi ipsi, modo jam explicato, omnino similis, aut aequalis. Hoc autem secundum impossibile esse sic demonstro. Nam ex ipsis terminis constare potest, quod recta linea, ex puncto A per X ulterius producta, unicam tantum sortiri potest in eo tali plano positionem, dum scilicet quaedam superaddita XF aeque omnino se habeat in laevam, et in dexteram partem praesuppositae AX, seu non magis in laevam, quam in dexteram ejusdem partem convergat. Non ergo alia erit continuatio AXM, quae rursum aeque omnino se habeat in laevam, et in dexteram partem ejusdem AX. Scilicet constat subsistere simul non posse; et quod continuatio AXF versus dexteram similis plane sit, aut aequalis sibi ipsi consideratae versus laevam; et quod alia quaedam continuatio AXM

end, that AXM with its fixed segment AX can again be so placed in that plane as to fit the other, AXF in so far as truly the prolongation AXF toward the right is precisely similar, or equal to the prolongation AXM toward the left, and moreover the prolongation AXF toward the left is precisely similar, or equal to the prolongation AXM toward the right.

But since the arc BC is greater than its part [79] MF, and in this portion MF in like way may be designated two other points with an interval less, without any certain limit; truly one of two things must happen in this approximation of the aforesaid points.

One is, if at length is attained one and the same intermediate point D, to which if XD is joined, such a prolongation AXD is obtained, as alone is such as to be wholly similar, or equal to itself (comparison made between the left and the right side).

The other is, if two such distinct points M, and F are found, to which XM, and XF being joined, two prolongations arise, one AXM, and the other AXF, of which each is, in the way just explained, wholly similar, or equal.

But this second I prove to be impossible thus. For from the very terms can be established, that a straight line produced from the point A on through X, can take in the plane only a single position, whilst obviously the superadded XF lies altogether equally toward the left, and toward the right side of the assumed AX, or deviates not more toward the left, than toward the right side of it. Therefore there will not be another prolongation AXM, which also lies altogether equally toward the left, and toward the right of this AX.

Obviously it holds that it cannot happen at the same time, both that the prolongation AXF toward the right is wholly similar, or equal to itself considered toward the left, and that another prolongation AXM toward the

versus laevam (quae, ex ipsa positione, minor sit continuatione AXF versus eandem laevam) aequalis iterum sit eidem continuationi versus dexteram, quae certe, ex ipsa rursum positione, major est praedicta continuatione AXF versus eandem dexteram.

Non ergo in eo arcu BC duo talia inveniri possunt puncta M, et F, ad quae junctae XM, et XF, duas exhibeant continuationes, unam AXM, et alteram AXF, qua-[80] rum utraque sit sibi ipsi, modo jam explicato, omnino similis, aut aequalis. Unde tandem consequitur incidi aliquando debere in unum, idemque punctum D, ad quod juncta XD talem exhibeat continuationem AXD, cui soli conveniat (facta comparatione inter laevam, ac dexteram partem) esse sibi ipsi omnino similem, aut aequalem. Quod erat hoc loco demonstrandum.

Dico tandem quinto: eam solam AXD fore lineam *rectam,* nimirum ex A per X *directe* continuatam in D. Quamvis enim ly *ex aequo,* in definitione lineae rectae, applicari primitus debeat punctis intermediis relate ad puncta ipsius extrema; unde utique jam elicuimus, *duas lineas rectas non claudere spatium*; intelligi tamen etiam debet de ejusdem rectae lineae continuatione *in directum.* Itaque ea sola XD (in eodem cum AX plano existens) dicetur esse continuatio *recta,* sive *in rectum* praedictae AX, quando ipsa neque in laevam, neque in dexteram illius partem convergat, sed utrinque *ex aequo* procedat; adeo ut nempe continuatio illa AXD versus laevam similis plane sit, aut aequalis eidem continuationi consideratae versus dexteram. Inde enim fiet, ut illi soli AXD conveniat non posse ab ea suscipi in eo tali plano aliam positionem sub illa immota AX; cum certe (ex jam demon-

left (which, from its very position, is less than the prolongation AXF toward the same left) again is equal to the same continuation toward the right, which truly, again from its very position, is greater than the aforesaid prolongation AXF toward the same right.

Therefore in the arc BC cannot be found two such points M, and F, that the joins XM, and XF, present two prolongations, one AXM, and the other AXF, of which [80] each is to itself, in the way just explained, wholly similar, or equal.

Whence at length follows, that somewhere must be attained one and the same point D, to which the join XD presents such a prolongation AXD, that to it alone belongs to be wholly similar, or equal to itself (comparison made between left, and right side).

Quod erat hoc loco demonstrandum.

At length I say fifthly: this AXD alone is a *straight* line, namely from A through X *directly* continued on to D.

For though the phrase *ex aequo,* * in the definition of the straight line, should primarily be applied to points intermediate in relation to its extreme points; whence in particular we have just deduced, *two straight lines do not inclose a space*; nevertheless it should also be understood of the *direct* prolongation of this straight line.

Therefore alone this AD (lying in the same plane with AX) is said to be the *straight* prolongation (or *in a straight*) of the aforesaid AX, when that deviates neither toward the left, nor toward the right side of it, but from each side proceeds *ex aequo*; so that the prolongation AXD is toward the left clearly similar, or equal to the same prolongation considered toward the right. For thence it will follow, that alone to AXD pertains that another position cannot be taken by it in the plane, while AX is fixed; when truly (from what

stratis) illae aliae AXB, et AXM, citra omnem suarum talium continuationum immutationem, suscipere possint sub eadem immota AX alias in eodem plano positiones, quales sunt ipsarum AXC, et AXF. Igitur illa sola AXD, cujus nempe continuatio XD tum in eodem cum ipsa AX plano existat, tum etiam aeque omnino se habeat in laevam, ac dexteram partem praedictae AX, est linea *recta* juxta explicatam definitionem, seu continuatio *in rectum* ejusdem praesuppositae rectae AX.

Ex quibus omnibus tandem constat evenire non posse, ut unum quodpiam sit commune segmentum duarum [81] rectarum. Quod erat demonstrandum.

COROLLARIUM.

Ex duobus praemissis Lemmatis tria opportune subnotare licet. Unum est: duas rectas, neque sub infinite parva inter ipsas distantia, claudere spatium posse. Ratio est, quia (prout in primo Lemmate) vel utraque illarum sub duobus illis communibus extremis punctis immotis revolvi posset ad novum situm occupandum, et sic (ex jam tradita lineae rectae definitione) neutra foret linea recta: vel una tantum in suo eodem situ persisteret, et sic illa sola recta linea foret. Quod autem nequeat utraque in eodem ipso situ persistere, dum aliquod concludant spatium, etiamsi infinite parvum, manifestum fiet consideranti posse faciem illius plani, in quo illae duae consistunt, converti de superna in infernam, manentibus caeteroquin in suo eodem loco duobus illis extremis punctis.

Alterum est: neque item ullam lineam rectam, in quantalibet ejusdem productione in directum, diffindi posse in duas, quamvis sub infinite parva intercapedine. Ratio est; quia (prout in praecedente Lemmate) conti-

has just now been proved) those others, AXB, and AXM, without any change of their prolongations, can, with the same fixed AX, take other positions in the same plane, such as AXC and AXF.

Therefore alone AXD, whose prolongation XD not only is in the same plane with AX, but also lies altogether in like manner toward the left, and the right side of the aforesaid AX, is a *straight* line in accordance with the discussed definition, or the prolongation *in a straight* of the assumed straight AX.

From all which finally is established as impossible, that one segment can be common to two [81] straight lines.

Quod erat demonstrandum.

COROLLARY.

From the two preceding lemmata three things may opportunely be noted.

One is: not even with an infinitely small distance between them can two straights inclose a space.

The reason is, because (just as in Lemma I) either each of them with the two common extreme points fixed can be revolved into occupying a new position, and so (from the definition of the straight line already given) neither will be a straight line: or only one remains in the same place, and so it alone is a straight line.

But that both cannot remain in the same place, while they inclose any space, even if infinitely little, will be manifest from considering that a face of the plane, in which the two are, can be converted from upper to lower, the two extreme points withal remaining in the same place.

Another is: nor moreover can any straight line, in any direct production of it, split into two, although with an interval infinitely small.

The reason is, because (just as in the preceding

nuatio in directum praesuppositae cujusdam simplicis rectae AX non alia esse intelligitur praeter unam XD, quae *ex aequo* utrinque procedat relate ad laevam, ac dexteram partem praedictae AX; ex quo utique fiat, ut sub ea immota AX non aliam ipsa immutata habere possit in eo plano positionem. Quod autem in eodem plano alia quaedam ad laevam decerni possit XM, infinite parum dissiliens ab ipsa XD, nihil suffragatur. Nam rursum alia item ad dexteram designari poterit XF, quae uniformiter infinite parum dissiliat ab eadem XD. Quare (prout in praecitato Lemma-[82]te) illa sola AXD erit linea recta a nobis definita.

Tertium tandem est: in hoc ipso secundo Lemmate censeri posse immediate demonstratam 1. undecimi; quod nempe ejusdem rectae nequeat pars una quidem in subjecto plano existere, et altera in sublimi.

LEMMA III.

Si duae rectae AB, CXD sibi invicem occurrant (fig. 39.) in aliquo ipsarum intermedio puncto X, non ibi se invicem contingent, sed una alteram ibidem secabit.

Demonstratur. Si enim fieri potest, tota CXD ad unam eandemque partem ipsius AB consistat. Jungatur AC. Non erit porro AC eadem cum ipsa veluti continuata AXC; quia caeterum (contra praecedens Lemma) duarum rectarum, unius AXC, et alterius praesuppositae DXC, unum idemque foret commune segmentum XC. Itaque jungatur BC. Non erit rursum haec BC continuatio ipsius BA usque in punctum C; ne duae rectae, una XAC, portio ipsius BAC, et altera XC spatium claudant, contra praemissum Lemma primum. Igitur ea BC vel se-

lemma) the direct prolongation of any assumed simple straight AX cannot be understood to be other than the one XD, which proceeds *ex aequo* on both sides in relation to the left, and right side of the aforesaid AX; from which assuredly follows, that with AX fixed it cannot, itself unchanged, have another position in this plane.

But that in the same plane a certain other XM can be designated to the left, splitting infinitely little from XD, nothing avails. For again another, XF, likewise to the right could be designated, which just so splits infinitely little from the same XD. Wherefore (as in the the lemma before cited) [82] alone AXD will be the straight line defined by us.

The third finally is: in this second lemma may be judged immediately demonstrated Eu. XI. 1; that of the same straight one part cannot be in a lower plane, and another in an upper.

LEMMA III.

If two straights AB, CXD meet each other (fig. 39) in any intermediate point X of theirs, they do not there touch each other, but one cuts the other there.

PROOF. For if that were possible, the whole CXD lies on one and the same side of AB. Join AC. Then AC will not be the same with AXC as if prolonged; because otherwise (against the preceding lemma) of two straights, one AXC, and the other the assumed DXC, there would be one and the same common segment XC.

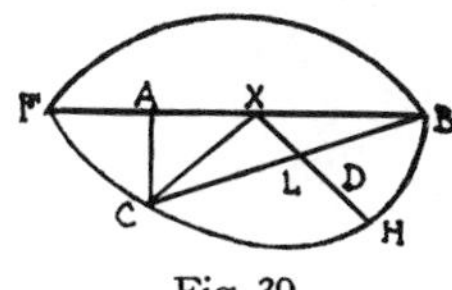

Fig. 39.

And so join BC. Again this BC will not be a prolongation of BA to the point C; lest two straights, one XAC, portion of this BAC, and the other XC inclose a space, against the preceding Lemma I.

cabit in aliquo puncto L ipsam XD, sive praesuppositam rectam DXC; et tunc rursum duae rectae lineae, una LC portio ipsius BC, et altera LXC portio praedictae DXC, spatium claudent; vel alterutrum extremum punctum sive A ipsius BA, sive D ipsius CXD, claudetur intra spatium comprehensum ipsis CX, XB, et alterutra vel BFC, vel BHC. At in utroque casu idem absurdum consequitur: Sive enim BA protracta per A occurrat ipsi BFC in aliquo puncto F; sive CXD protracta per D occurrat ipsi BHC in aliquo puncto H: in idem semper absurdum incidimus, quod duae rectae spatium claudant; nimirum aut [83] recta BF portio ipsius BFC una cum altera BAF; aut recta HC, portio ipsius BHC, una cum altera praesupposita recta continuata CXDH.

Porro idem, aut majus absurdum consequitur, si illa BA protracta per A occurrat in aliquo puncto vel ipsi CX, vel sibi ipsi in aliquo puncto suae portionis XB. Atque id similiter valet, si altera CXD protracta per D occurrat in aliquo puncto vel ipsi XB, vel sibi ipsi in aliquo puncto suae portionis CX.

Itaque constat, quod duae rectae AB, CXD sibi invicem occurrentes in aliquo ipsarum intermedio puncto X, non ibi se invicem contingent, sed una alteram ibidem secabit. Quod erat etc.

LEMMA IV.

Omnis diameter dividit bifariam suum circulum, ejusque circumferentiam.

Demonstratur. Esto circulus (recole fig. 23.) MDHNKM, cujus centrum A, et diameter MN. Intelligatur illius circuli portio MNKM ita revolvi circa immota puncta M, et N, ut tandem accommodetur, seu coaptetur reliquae portioni MNHDM. Constat primo totam dia-

Therefore this BC either will cut XD (or the as-
umed straight DXC) in some point L; and then again
wo straight lines, one LC, portion of this BC, and the
ther LXC, portion of the aforesaid DXC, inclose a
pace; or one of the extreme points whether A of BA,
r D of CXD, is inclosed within the space bounded by
X, XB, and either BFC, or BHC.

But in either case the same absurdity follows: For
vhether BA produced through A strikes BFC in a point
; or CXD produced through D strikes BHC in a point
I; always we come upon the same absurdity, that two
traights inclose a space; forsooth either the straight[83]
F portion of BFC together with the other BAF; or the
traight HC, portion of BHC, together with the other
ssumed straight prolonged CXDH.

Furthermore the same, or a greater absurdity fol-
ows, if BA produced through A meets in any point
ither CX, or its own self in any point of its portion XB.

And this likewise holds, if the other CXD produced
hrough D meets in any point either XB, or its own self
n any point of its portion CX.

Therefore is established, that two straights AB, CXD
neeting each other in any intermediate point X of theirs,
o not there touch each other, but one will cut the other
here.

Quod erat etc.

LEMMA IV.

very diameter bisects its circle, and the circumference of it.

Proof. Let there be a circle (recall fig. 23) MDH-
IKM, A its center, and MN a diameter. Of this circle
he portion MNKM is thought so to revolve about the
xed points M, and N, that at length it is superimposed
pon, or applied to the remaining portion MNHDM.

metrum MAN quoad omnia ipsius puncta in eodem situ esse mansuram: ne duae rectae lineae (contra praecedens Lemma primum) spatium claudant. Constat secundo nullum punctum K circumferentiae NKM casurum vel intra, vel extra superficiem clausam diametro MAN, et altera circumferentia NHDM; ne scilicet contra naturam circuli, unus radius v. g. AK minor sit, aut major altero ejusdem circuli radio v. g. AH. Constat tertio quemlibet radium MA continuari unice posse in rectum per[84] alterum quendam radium AN, ne (contra praecedens Lemma secundum) duae suppositae rectae lineae, ut puta MAN, MAH, unum idemque commune habeant segmentum MA. Constat quarto (ex proxime antecedente Lemmate) omnes cujusvis circuli diametros se invicem in centro secare, et ex nota natura circuli bifariam.

Ex quibus omnibus constare potest, quod diameter MAN tum dividit exactissime suum circulum, ejusque circumferentiam in duas aequales partes, tum etiam assumi universim potest pro qualibet ejusdem circuli diametro. Quod erat etc.

SCHOLION.

Hanc eandem veritatem demonstratam leges apud Clavium a Thalete Milesio, sed fortasse non exhausta omni qualibet objectione.

LEMMA V.

Inter angulos rectilineos omnes anguli recti sunt invicem exactissime aequales, sine ullo defectu etiam infinite parvo.

Demonstratur. Angulum inter rectilineos rectum definit Euclides: *qui est aequalis suo deinceps.* Non hunc

It is certain first that as to all its points the whole diameter MAN will remain in the same place; lest two straight lines (against the preceding Lemma I) inclose a space.

It is certain secondly that no point K of the circumference NKM will fall either within, or without the surface inclosed by the diameter MAN, and the other circumference NHDM; lest obviously against the nature of the circle, one radius, for example AK, be less, or greater than another radius of the same circle, for example AH.

It is certain thirdly that any radius MA can alone be prolonged in a straight line by [84] a certain other radius AN, lest (against the preceding Lemma II) two lines assumed straight, as suppose MAN, MAH, should have one and the same common segment MA.

It is certain fourthly (from the immediately preceding lemma) that all the diameters of the circle cut one another in the center, and from the known nature of the circle bisect.

From all which can be established, that not only the diameter MAN most exactly divides its circle, and the circumference of it into two equal parts, but also that this may be assumed universally for any diameter of this circle. Quod erat etc.

SCHOLION.

We read in Clavius that this truth was demonstrated by Thales of Miletus, but perhaps not to the exhaustion of every objection.

LEMMA V.

Among rectilinear angles, all right angles are exactly equal to one another, without any deviation even infinitely small.

Proof. Euclid defines a rectilinear angle as right: *which is equal to its adjacent.*✲ This he does not postu-

postulat ipse sibi concedi, sed problematice demonstrat in sua Prop. XI. Libri primi. Ibi enim ex dato in recta BC quolibet puncto A (fig. 40.) docet excitare perpendicularem AD, ad quam anguli DAB, DAC sint invicem aequales. Porro illos duos angulos esse invicem exactissime aequales, sine ullo defectu etiam infinite parvo constare potest ex Corollario post duo priora praemissa Lemmata [85] si nempe ipsae AB, AC designatae sint exactissime aequales.

Sed aliqua oriri potest dubitatio, si duo alii ad quandam alteram FM recti anguli LHF, LHM (fig. 41.) conferantur cum praedictis rectis angulis DAB, DAC. Itaque HL aequalis sit ipsi AD, ac rursum posterior integra Figura ita intelligatur superponi priori, ut punctum H cadat super punctum A, et punctum L super punctum D. Jam sic progredior. Et prima quidem (ex praecedente Lemmate) ipsa FHM non praecise continget alteram BC in eo puncto A. Ergo vel adamussim procurret super illa BC, vel eandem ita secabit, ut unum ejus punctum extremum v. g. F cadat supra, et alterum M deorsum. Si primum: jam clare habemus exactissimam inter omnes rectilineos angulos rectos aequalitatem intentam. At non secundum; quia sic angulus LHF, hoc est DAF, minor foret angulo DAB, ejusque supposito exactissime aequali

late as conceded to him, but demonstrates through a problem in his Bk. I. P. 11. For there he teaches from any given point A (fig. 40) in the straight BC to erect a perpendicular AD at which the angles DAB, DAC are equal to each other.

Moreover that those two angles are precisely equal to each other, without any difference even infinitely small, follows from the corollary after the first two premised lemmata, [85] if AB, AC are taken exactly equal.

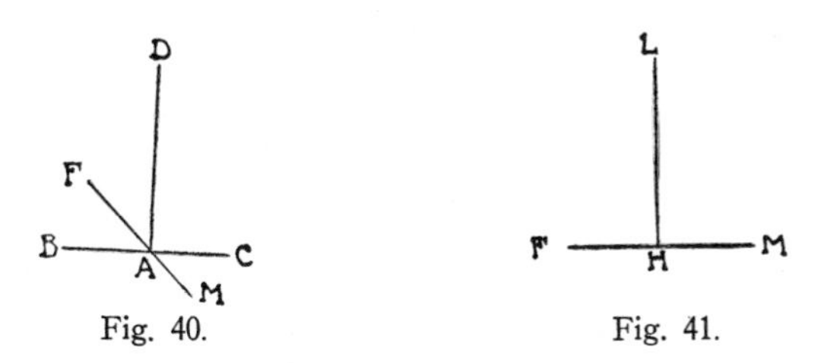

Fig. 40. Fig. 41.

But some doubt may arise, if two other right angles LHF, LHM (fig. 41) at any other straight FM are compared with the aforesaid right angles DAB, DAC. ☼

Therefore let HL be equal to AD, and then the whole latter figure is thought to be superposed upon the former so, that point H falls upon point A, and point L upon point D.

Now I proceed thus.

And first indeed (from a preceding lemma [III]) this FHM does not exactly touch the other BC in the point A. Therefore either it runs forward precisely upon BC, or will cut it so that one of its end points for example F falls above, and the other M below.

In the first case: now clearly we have the exact equality asserted between all rectilinear right angles.

But not in the second; because thus the angle LHF, here it is DAF, will be less than the angle DAB, and its

DAC, et sic multo minor angulo DAM, sive LHM; contra hypothesin. Deinde vero nihil suffragatur, quod angulus DAF infinite parum deficiat ab angulo DAB, sive ejus exactissime aequali DAC, qui rursum solum infinite parum superetur ab angulo DAM. Nam semper angulus DAF, sive LHF, non erit exactissime aequalis angulo DAM, sive LHM, contra hypothesin.

Itaque constat omnes rectilineos angulos rectos esse invicem exactissime aequales, sine ullo defectu etiam infinite parvo. Quod etc.

COROLLARIUM.

Inde autem fit, ut quae ex uno dato cujusvis rectae lineae puncto perpendiculariter in aliquo plano ad eandem educitur, ipsa sit in eo tali plano unica exactissime linea recta, nec potens diffindi in duas. [86]

Post quinque praemissa Lemmata, eorumque Corollaria, progredi jam debeo ad demonstrandum principale assumptum contra hypothesin anguli acuti.

Ubi statuere possum, tanquam per se notum, non minus repugnare, quod duae rectae lineae (sive ad finitam, sive ad infinitam earundem productionem) in unam tandem, eandemque rectam lineam coeant; quam quod una eademque linea recta (sive ad finitam, sive ad infinitam ejusdem continuationem) in duas rectas lineas diffindatur, contra praecedens Lemma secundum, ejusque Corollarium. Quoniam ergo naturae lineae rectae (ex praecedente Corol. proximi Lemmatis) oppositum itidem est, quod duae rectae lineae ad unum, idemque punctum cujusdam tertiae rectae, perpendiculares ipsi sint in eodem communi plano; agnoscere oportet tanquam absolute falsam, quia repugnantem naturae praedictae, hypothesin anguli acuti, juxta quam duae illae AX, BX (fig. 33.) in uno

supposed exact equal DAC, and thus much less than the angle DAM, or LHM; contrary to the hypothesis.

Then it helps nothing that angle DAF differ infinitely little from angle DAB, or its exact equal DAC, which again would exceed only infinitely little the angle DAM. For always angle DAF, or LHF, will not be exactly equal to angle DAM, or LHM, against the hypothesis.

Therefore is established that all rectilinear right angles are exactly equal to one another, without any difference even infinitely small.

Quod erat etc.

COROLLARY.

Thence follows, that the straight line erected from a given point of any straight perpendicularly to it in a plane, is, in such plane, wholly unique, nor can it split in two. [86]

After the five premised lemmata, and their corollaries, I must now go on to proof of the principal objection against the hypothesis of acute angle.

Here I may set up, as known *per se,* it is not less contradictory, that two straight lines (whether at a finite, or at an infinite prolongation of them) at length run together into one and the same straight line, than that one and the same straight line (whether at a finite, or at an infinite prolongation of it) splits into two straight lines, against the preceding Lemma II, and its corollary.

Since therefore it is in like manner opposed to the nature of the straight line (from the preceding corollary to the last lemma), that two straight lines at one and the same point of a third straight, be perpendicular to this in the same common plane; it is proper to recognize as absolutely false, because repugnant to the aforesaid nature, the hypothesis of acute angle, according to which those two AX, BX (fig. 33) in one and the same com-

eodemque communi puncto X perpendiculares esse debe-
rent cuidam tertiae rectae, quae in eodem cum ipsis plano existeret. Hoc autem erat principale demonstrandum.

SCHOLION.

Atque his subsistere tutus possem. Sed nullum non movere lapidem volo, ut inimicam anguli acuti hypothesim, a primis usque radicibus revulsam, sibi ipsi repugnantem ostendam. Iste autem erit consequentium hujus Libri Theorematum unicus scopus. [87]

mon point X must be perpendicular to a third straight, which is in the same plane with them. ✡

Hoc autem erat principale demonstrandum.

SCHOLION.

And here I might safely stop. But I do not wish to leave any stone unturned, that I may show the hostile hypothesis of acute angle, torn out by the very roots, contradictory to itself.

However this will be the single aim of the subsequent theorems of this Book. [87]

LIBRI PRIMI PARS ALTERA.

IN QUA IDEM PRONUNCIATUM EUCLIDAEUM CONTRA HYPOTHESIN ANGULI ACUTI REDARGUTIVE DEMONSTRATUR.

PROPOSITIO XXXIV.

In qua expenditur curva quaedam enascens ex hypothesi anguli acuti.

Recta CD jungat aequalia perpendicula AC, BD cuidam rectae AB insistentia. Tum divisis bifariam in punctis M, et H (fig. 42.) ipsis AB, CD, jungatur MH (ex 2. hujus) utrique perpendicularis. Rursum in hac hypothesi supponuntur acuti anguli ad junctam CD. Quare in quadrilatero AMHC erit MH (ex Cor. I. post 3. hujus) minor ipsa AC. Hinc autem; si in MH protracta sumas MK aequalem ipsi AC; puncta C, K, D, spectabunt ad curvam hic expensam. Deinde anguli ad junctam CK erunt et ipsi (ex 7. hujus) acuti. Igitur juncta LX, quae bifariam, atque ideo (ex 2. hujus) ad angulos rectos dividat ipsas AM, CK, erit similiter (ex Cor. I. post 3 hujus) minor eadem AC. Quapropter; si in LX protracta sumas LF aequalem ipsi AC, aut MK; etiam punctum F spectabit ad eam curvam. Praeterea jungens CF, et FK invenies similiter duo alia puncta ad eandem curvam spectantia. Atque ita semper. Quod autem dico pro inve-

PART II.

IN WHICH THE SAME EUCLIDEAN POSTULATE IS DEMONSTRATED AGAINST THE HYPOTHESIS OF ACUTE ANGLE BY REFUTING THIS.

PROPOSITION XXXIV.

In which is investigated a certain curve arising from the hypothesis of acute angle.[1]

Let the straight CD join equal perpendiculars AC, BD standing upon a certain straight AB. Then AB, CD being bisected in the points M and H (fig. 42), MH is joined perpendicular (by P. II.) to each. Again in this hypothesis the angles at the join CD are supposed acute. Therefore in the quadrilateral AMHC (by Cor. I. to P III.) MH will be less than AC. Hence now, if in MH produced MK be taken equal to AC, the points C, K, D pertain to the curve here investigated. Then the angles at the join CK will be themselves acute (by P. VII.).

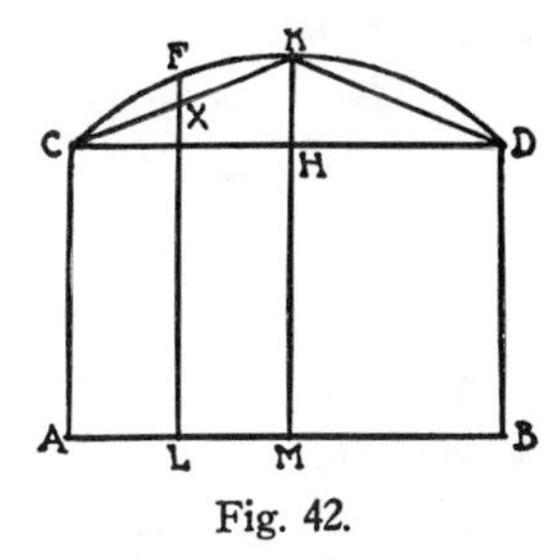

Fig. 42.

Therefore the join LX, which bisects, and therefore (by P. II.), is at right angles to AM, CK, will be likewise (by Cor. I. to P. III.) less than AC. Wherefore, if in LX produced we assume LF equal to AC or MK, the point F also will pertain to this curve. Further, joining CF, and FK we find likewise two other points pertaining to the same curve. And so on forever.

[1] In the hypothesis of acute angle an equidistant of a straight has its chords between it and the straight.

niendis punctis inter puncta C, et K, idem etiam uniformiter valet pro inveniendis punctis inter puncta K, et D, scilicet cur-[88]va CKD, enascens ex hypothesi anguli acuti, est linea jungens extremitates omnium aequalium perpendiculorum super eadem basi versus eandem partem erectorum, quae utique venire possunt sub nomine rectarum ordinatim applicatarum; est, inquam, linea ejusmodi, quae propter ipsam, ex qua nascitur, hypothesim anguli acuti, semper est cava versus partes contrapositae basis AB. Quod quidem hoc loco declarandum, ac demonstrandum a nobis erat.

PROPOSITIO XXXV.

Si ex quolibet puncto L basis AB ordinatim applicetur ad eam curvam CKD recta LF: Dico rectam NFX perpendicularem ipsi LF cadere totam ex utraque parte debere versus partes convexas ejusdem curvae, atque ideo eam fore ejusdem curvae tangentem.

Demonstratur. Si enim fieri potest, cadat quoddam punctum X (fig. 43.) ipsius NFX intra cavitatem ejusdem curvae. Demittatur ex puncto X ad basim AB perpendicularis XP, quae protracta per X occurrat curvae in quodam puncto R. Jam sic. In quadrilatero LFXP non erit angulus in puncto X aut rectus, aut obtusus: Caeterum (ex 5. et 6. hujus) destrueretur praesens hypothesis anguli acuti. Ergo praedictus angulus erit acutus. Quare erit PX (ex Cor. I. post 3. hujus) et sic multo magis PR major ipsa LF. Hoc autem absurdum est (ex praecedente) contra naturam istius curvae. Itaque illa NF pro-

But what I say for finding points between the points C and K, the same also holds good uniformly for finding points between the points K and D; of course the curve [88] CKD, arising from the hypothesis of acute angle, is the line joining the extremities of all equal perpendiculars erected upon the same base toward the same part, which assuredly can come under the name ordinates; it is, I may say, a line of such sort, that on account of the hypothesis of acute angle, from which it arises, it always is concave toward the parts of the opposite base AB.

Quod quidem hoc loco declarandum, ac demonstrandum a nobis erat.

PROPOSITION XXXV.

If from any point L of the base AB the ordinate LF is drawn to this curve CKD: I say the straight NFX perpendicular to LF must on both sides fall wholly toward the convex parts of this curve, and therefore it will be tangent to this curve.

PROOF. For if possible, let a certain point X (fig. 43) of NFX fall within the cavity of this curve. Let fall from the point X to the base AB the perpendicular XP, which prolonged through X meets the curve in a certain point R. Now thus. In the quadrilateral LFXP the angle at the point X will be neither right nor obtuse: else (P. V. and P. VI.) would be destroyed the present hypothesis of acute angle.

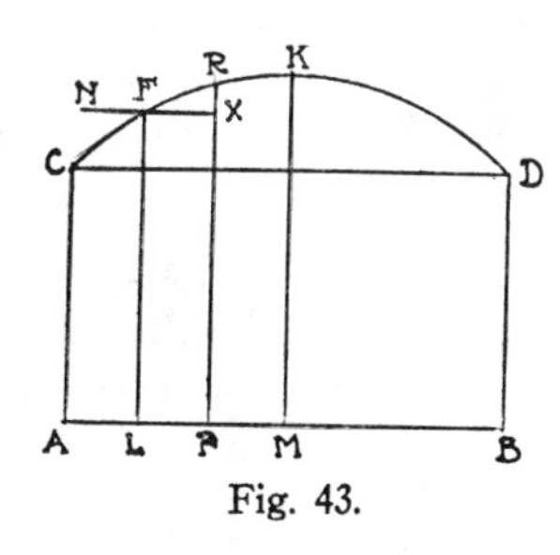

Fig. 43.

Therefore the aforesaid angle will be acute. Wherefore (from Cor. I. to P. III.) PX and so much more PR will be greater than LF. But this is absurd (from the preceding) against the nature of this curve.

tracta cadere tota debet versus partes convexas, atque ideo ipsa erit ejusdem curvae tangens. Quod erat demonstrandum. [89]

PROPOSITIO XXXVI.

Si recta quaepiam XF (fig. 44.) acutum angulum efficiat cum quavis ordinata LF, non cadet punctum X extra cavitatem curvae, nisi prius ipsa XF in aliquo puncto O curvam secuerit.

Demonstratur. Constat sumi posse in ipsa XF punctum quoddam X adeo vicinum ipsi puncto F, ut juncta LX prius curvam secet in aliquo puncto S: caeterum ipsa XF vel non cadet tota extra cavitatem curvae, et sic habemus intentum; vel adeo non efficiet cum FL angulum acutum, ut magis censenda jam sit in unicam rectam cum altera LF coire. Itaque ex puncto S demittatur ad basim AB perpendicularis SP. Erit haec (ex 34. hujus) aequalis ipsi LF. Est autem SP (ex 19. primi) minor ipsa LS. Ergo etiam LF minor est eadem LS, ac propterea multo minor ipsa LX. Hinc in triangulo LXF acutus erit angulus in puncto X, quia minor (ex 18. primi) angulo LFX supposito acuto. Jam demittatur ad FX perpendicularis LT. Cadet haec (propter 17. primi) ad partes utriusque anguli acuti. Quare punctum T jacebit inter puncta X, et F. Deinde ex puncto T demittatur ad basim AB perpendicularis TQ. Erit LF (propter angulum

So NF produced must fall wholly toward the convex parts, and so it will be tangent to this curve.

Quod erat demonstrandum. [89]

PROPOSITION XXXVI.

If any straight XF (fig. 44) makes an acute angle with any ordinate LF, the point X does not fall without the cavity of the curve, unless previously XF has cut the curve in some point O.

PROOF. It is sure that some point X may be assumed in XF so near to the point F, that the join LX previously cuts the curve in some point S: otherwise XF either does not fall wholly without the cavity of the curve, and so we have our assertion; or so far is it from making with FL an acute angle, that now rather it must be supposed to combine with LF in one straight.

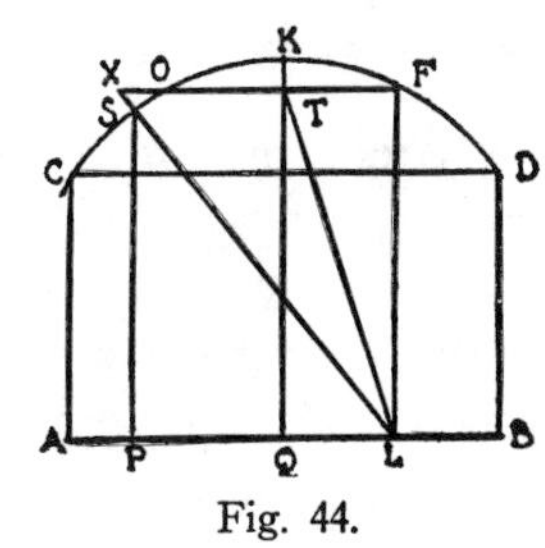

Fig. 44.

Accordingly from the point S let fall to the base AB the perpendicular SP. This will be (from P. XXXIV.) equal to LF.

But SP is (from Eu. I. 19) less than LS. Therefore also LF is less than LS, and consequently much less than LX. Hence in triangle LXF the angle at point X will be acute, because less (from Eu. I. 18) than the angle LFX supposed acute.

Now let fall to FX the perpendicular LT. This will fall (because of Eu. I. 17) toward the parts of each acute angle. Wherefore point T will lie between points X, and F.

Then from the point T let fall to the base AB the perpendicular TQ. LF will be (because of the right

rectum in T) major ipsa LT, et haec (propter angulum rectum in Q) major altera QT. Igitur LF multo major erit ipsa QT. Hinc autem; si in QT protracta sumatur QK aequalis ipsi LF; punctum K (ex 34. hujus) ad praesentem curvam spectabit, cadetque idcirco punctum T intra cavitatem ejusdem curvae. Non ergo recta FT, quae secat duas rectas QK, et LT in T, promoveri potest ad secandam protractam LS in puncto X, constituto extra cavitatem praesentis curvae, nisi prius ipsa protracta FT secet in aliquo puncto O portionem ejusdem curvae inter puncta S, et K [90] constitutam. Hoc autem erat demonstrandum.

COROLLARIUM.

Atque hinc manifeste liquet, inter tangentem hujus curvae, et ipsam curvam locari non posse quandam rectam, quae tota ad hanc, vel illam tangentis partem extra curvae cavitatem cadat; quandoquidem recta sic locata efficere debet (ex praecedente) angulum acutum cum demissa ex puncto contactus ad contrapositam basim perpendiculari.

PROPOSITIO XXXVII.

Curva CKD, ex hypothesi anguli acuti enascens, aequalis esse deberet contrapositae basi AB.

Ante demonstrationem praemitto sequens axioma.

Si duae lineae bifariam dividantur, tum earum medietates, ac rursum quadrantes bifariam, atque ita in infinitum uniformiter procedatur; certo argumento erit, duas istas lineas esse inter se aequales, quoties in ista uniformi in infinitum divisione comperiatur, seu demonstretur, de-

angle at T) greater than LT, and this (because of the right angle at Q) will be greater than QT. Therefore LF will be much greater than QT. But hence; if in QT produced QK is taken equal to LF; the point K (from P. XXXIV.) will pertain to the present curve, and therefore point T falls within the cavity of this curve.

Therefore the straight FT, which cuts the two straights QK, and LT in T, cannot be extended to cut LS prolonged in the point X, situated without the cavity of the present curve, unless previously the prolonged FT cuts in some point O the portion of this curve situated between the points S, and K. [90]

Hoc autem erat demonstrandum.

COROLLARY.

✲ And hence it is clear that between the tangent of this curve, and the curve itself cannot be placed any straight, which, on one or the other side of the tangent wholly falls without the cavity of the curve; since a straight so located must (from the preceding) make an acute angle with the perpendicular let fall from the point of contact to the opposite base.

PROPOSITION XXXVII.

The curve CKD, arising from the hypothesis of acute angle, must be equal to the opposite base AB.

Before the demonstration I premise the following axiom.

If two lines be bisected, then their halves, and again their quarters bisected, and so the process be continued uniformly *in infinitum*; it will be safe to argue, those two lines are equal to each other, as often as is ascertained, or demonstrated in that uniform division *in infinitum,* that at length must be attained two of their mutually

veniri tandem debere ad duas illarum sibi invicem respondentes partes, quas constet esse inter se aequales.

Jam demonstratur propositum. Intelligantur erecta ex basi AB ad eam curvam CKD (fig. 45.) quotvis perpendicula NF, LF, PF, MK, TF, VF, IF; sintque aequales in ipsa basi AB portiones AN, NL, LP, PM, MT, TV, VI, IB.

Constat primo angulum ipsius AC cum ea curva aequalem fore singulis hinc inde ad puncta F, sive ad punctum K, aut punctum D, praedictarum perpendicularium angulis cum eadem curva. Si enim mistum quadrilate-[91]rum ANFC superponi intelligatur misto quadrilatero NLFF, constituta basi AN super aequali basi NL, cadet AC super NF, et NF super LF, propter aequales angulos rectos ad puncta A, N, L. Deinde propter aequalitatem rectarum (ex 34. hujus) AC, NF, LF, cadet punctum C super punctum F ipsius NF, et hoc super alterum punctum F ipsius LF. Praeterea curva CF congruet adamussim ipsi curvae FF: si enim una illarum, ut CF introrsum, aut extrorsum cadat; sumpto quolibet puncto Q inter puncta N, et L, ductaque perpendiculari secante unam curvam in X, et alteram in S, aequales forent (ex nota hujus curvae natura) ipsae QX, QS, quod est absurdum. Idem valebit, si in dicta superpositione maneat in suo situ recta NF, et recta AC cadat super LF. Rur-

corresponding parts, of which it is certain they are equal to each other.

Now for the proof of the proposition.

Suppose erected from the base AB to the curve CKD (fig. 45) indefinitely many perpendiculars NF, LF, PF, MK, TF, VF, IF; and on the base AB take as equal the portions AN, NL, LP, PM, MT, TV, VI, IB.

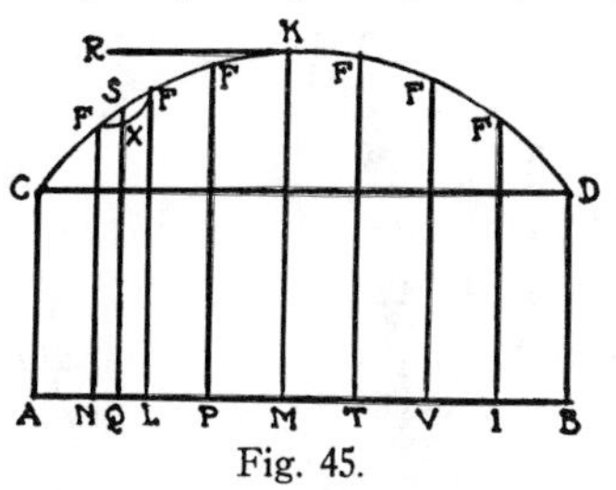

Fig. 45.

First is certain the angle of AC with the curve will be equal to each of the angles of the aforesaid perpendiculars with the curve on either side at the points F, or at the point K, or at the point D. For if the mixed quadrilateral [91] ANFC is supposed to be superposed upon the mixed quadrilateral NLFF, the base AN lying upon the equal base NL, AC falls upon NF, and NF upon LF, because of the equal right angles at the points A, N, L. Then because of the equality (from P XXXIV.) of the straights AC, NF, LF, the point C falls upon point F of NF, and this upon the other point F of LF.

Moreover the curve CF exactly fits the curve FF: for if one of these, as CF fell within or without; any point Q being assumed between points N, and L, and the perpendicular being drawn cutting one curve in X, and the other in S, QX, QS would be equal (from the known nature of this curve), which is absurd. The same will hold, if in the said superposition the straight NF remains in its place, and the straight AC falls upon LF. Again

sum idem valebit, si idem quadrilaterum mistum ANFC utrovis modo superponi intelligatur cuivis reliquorum quadrilaterorum usque ad ipsum inclusive postremum quadrilaterum BDFI. Itaque angulus ipsius AC cum ea curva aequalis est singulis hinc inde ad puncta F, sive ad punctum K, aut punctum D, praedictarum perpendicularium angulis cum eadem curva.

Constat hinc secundo aequales adamussim inter se esse portiones ipsius curvae ab istis perpendicularibus hinc inde abscissas.

Si ergo basis AB divisa sit bifariam in M, et medietas AM bifariam in L; tum quadrans LM bifariam in P; atque ita in infinitum, procedendo semper versus partes puncti M; constabit tertio, etiam curvam CKD bifariam dividi in K a perpendiculari MK, medietatem CK bifariam itidem dividi in F a perpendiculari LF, quadrantem FK bifariam in F a perpendiculari PF; atque ita in infinitum, procedendo semper uniformiter versus partes ipsius puncti K.

Quoniam vero in ista basis AB in infinitum divisione [92] considerare possumus rem devenisse ad portionem ipsius AB infinite parvam, quae nempe exhibeatur per latitudinem infinite parvam perpendicularis MK, constabit quarto (ex praemisso axiomate) aequalitas intenta totius basis AB cum tota curva CKD, dum alias ostendam portionem infinite parvam abscissam ex basi AB a perpendiculari MK aequalem esse adamussim portioni infinite parvae, quam eadem perpendicularis abscindit ex curva CKD. Et hoc quidem postremum sic demonstro.

Nam RK perpendicularis ipsi KM tanget (ex 35. hujus) curvam in K, atque ita eandem tanget in K, ut inter ipsam tangentem (ex Cor. post 36. hujus) et curvam, ex neutra parte locari possit recta, quae ipsam curvam non

the same will hold, if the same mixed quadrilateral ANFC in either mode is supposed to be superposed to any of the remaining quadrilaterals even to the last quadrilateral BDFI inclusive.

Therefore the angle of AC with the curve is equal to either of the angles with this curve of the aforesaid perpendiculars on either side at the points F, or at the point K, or point D.

Hence follows secondly that the portions of the curve cut off on each side by these perpendiculars are exactly equal to one another.

If therefore the base AB be bisected in M, and the half AM bisected in L; then the quarter LM bisected in P; and so *in infinitum,* proceeding always toward the parts of the point M; it will follow thirdly, also the curve CKD is bisected in K by the perpendicular MK, the half CK in like manner bisected in F by the perpendicular LF, the quarter FK bisected in F by the perpendicular PF; and so *in infinitum,* proceeding always uniformly toward the parts of the point K.

But since in this division of the base AB *in infinitum* we may [92] consider the thing to have arrived at a portion of AB infinitely small, which obviously may be exhibited by the infinitely small breadth of the perpendicular MK, fourthly (from the premised axiom) will follow the asserted equality of the whole base AB with the whole curve CKD, if only I now can show the infinitely small portion cut off from the base AB by the perpendicular MK to be exactly equal to the infinitely small portion, which the same perpendicular cuts off from the curve CKD.

And this last I thus demonstrate.

For RK perpendicular to KM touches (from P. XXXV.) the curve at K, and touches this in K so, that between the tangent (from Cor. to P. XXXVI.) and the curve from neither side can be placed a straight, which

secet. Igitur infinitesima K, spectans ad curvam, aequalis omnino erit infinitesimae K spectanti ad tangentem. Constat autem infinitesimam K spectantem ad tangentem, nec majorem, nec minorem, sed omnino aequalem esse infinitesimae M spectanti ad basim AB; quia nempe recta illa MK intelligi potest descripta ex fluxu semper ex aequo ejusdem puncti M usque ad eam summitatem K.

Quare (juxta praemissum axioma) curva CKD, ex hypothesi anguli acuti enascens, aequalis esse deberet contrapositae basi AB. Quod erat demonstrandum.

SCHOLION I.

Sed forte minus evidens cuipiam videbitur enunciata exactissima aequalitas inter illas infinitesimas M, et K. Quare ad avertendum hunc scrupulum sic rursum procedo. Cuidam rectae AB insistant ad rectos angulos in eodem plano (fig. 48.) duae rectae aequales AC, BD. Rursum in eodem plano intelligatur existere circulus BLDH, cujus diameter BD; sitque semicircumferentia BLD aequa-[93]lis praedictae AB. Praeterea idem circulus ita in eo plano revolvi concipiatur super ea recta AB, ut motu semper continuo, et aequabili perficiat, seu describat suae ipsius semicircumferentiae punctis praedictam BA; quousque nempe punctum D, ad illam semi-

does not cut the curve. Therefore infinitesimal K, regarding the curve, will be wholly equal to infinitesimal K regarding the tangent. But it is certain the infinitesimal K regarding the tangent is neither greater nor less than, but exactly equal to the infinitesimal M regarding the base AB; because obviously the straight MK may be supposed described by the flow always uniform of the point M up to the summit K.

Wherefore (according to the premised axiom) the curve CKD, born of the hypothesis of acute angle should be equal to the opposite base AB.

Quod erat demonstrandum.

SCHOLION I. ✡

But perchance to some one will seem by no means evident the enunciated exact equality between the infinitesimals M, and K. ✡Wherefore to remove this scruple I again proceed thus.

To a certain straight AB let two equal straights AC, BD (fig. 48) stand at right angles in the same plane.

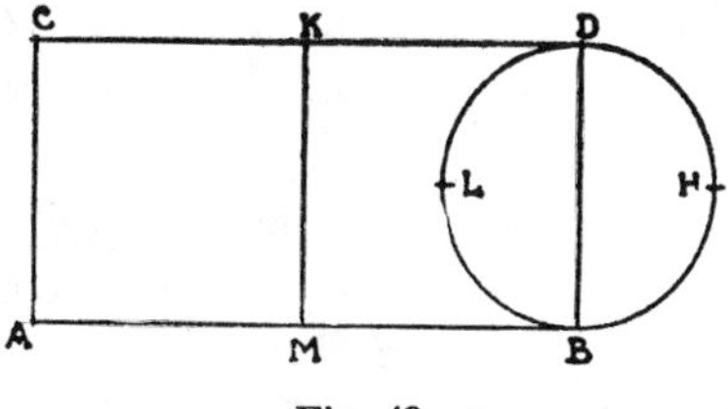

Fig. 48.

Again in the same plane suppose there is a circle BLDH, whose diameter is BD; and let the semicircumference BLD be equal [93] to the aforesaid AB. Further let the same circle be conceived so to be revolved in that plane upon the straight AB that with motion always continuous and uniform it achieves or describes with the points of its semicircumference the aforesaid BA, until

circumferentiam spectans, perveniat ad congruendum ipsi puncto A, ita ut propterea punctum B, ejusdem semicircumferentiae alterum extremum punctum, deveniat ad congruendum illi puncto C.

His stantibus; si in semicircumferentia BLD designetur quodvis punctum L, cui in descripta recta linea BA correspondeat punctum M, ex quo in eo tali plano educatur perpendicularis MK, aequalis ipsi BD: Dico illud punctum K fore ipsum punctum H diametraliter oppositum illi puncto L. Nam ibi in puncto M, sive L recta AB continget praedictum circulum. Igitur MK eidem AB perpendicularis transibit (ex 19. tertii, quae utique independens est ab Axiomate controverso) per centrum ejusdem circuli. Quare; ubi punctum L in ea tali circuli BLDH revolutione perveniat ad congruendum cum puncto M ipsius AB, etiam punctum H, diametraliter oppositum praedicto puncto L, incidet in punctum K illius MK.

Porro constat idem similiter valere de reliquis punctis semicircumferentiae BLD, et horum diametraliter correlativis in altera semicircumferentia BHD. Quare linea, eo tali modo successive descripta a punctis semicircumferentiae BHD, erit illa eadem jam expensa DKC, quae nempe suis omnibus punctis aequidistet ab illa recta BA; sitque idcirco (juxta hypothesin anguli acuti) semper cava versus partes ejusdem AB.

Inde autem fit, ut punctum M in ea BA censendum sit exactissime aequale puncto K in altera DKC, propter[94] omnimodam istorum aequalitatem cum punctis L, et H diametraliter oppositis in eo circulo BLDH. Quare; cum idem valeat de omnibus punctis descriptae rectae BA, si conferantur cum aliis uniformiter contrapositis in praedicta supposita curva DKC; consequens plane est, ut ipsa

indeed point D pertaining to that semicircumference comes to congruence with point A, so that moreover point B, the other extreme point of the same semicircumference comes to congruence with point C.

This abiding; if in the semicircumference BLD is designated any point L, to which in the described straight line BA corresponds point M, from which in that plane is erected the perpendicular MK, equal to BD: I say that point K will be the point H diametrically opposite the point L.

For there in the point M, or L the straight AB touches the aforesaid circle. Therefore MK perpendicular to AB will go (from Eu. III. 19,* which is assuredly independent of the controverted axiom) through the center of the same circle. Wherefore; where point L in that revolution of the circle BLDH comes to congruence with the point M of AB, also point H, diametrically opposite the aforesaid point L, falls upon point K of MK.

Furthermore it is certain the same holds in like manner of the remaining points of the semicircumference BLD, and of those diametrically correlative in the other semicircumference BHD. Wherefore the line, in that way successively described by the points of the semicircumference BHD, will be the already considered DKC, which in all its points is equidistant from the straight BA; and which therefore (in accordance with the hypothesis of acute angle) is always concave toward the side of AB.

But thence follows, that the point M in BA may be considered exactly equal to point K in DKC, because of [94] their equality in every way with the points L, and H diametrically opposite in the circle BLDH.

Wherefore; since the same holds of all points of the described straight BA, if they be compared with the other uniformly opposite in the aforesaid assumed curve DKC; the consequence evidently is, that this curve, born of the

talis curva, ex hypothesi anguli acuti enascens, censenda sit aequalis contrapositae basi AB. Atque id est, quod nova hac methodo iterum demonstrandum susceperam.

SCHOLION II.

Rursum vero: quoniam recta BA intelligitur successive descripta a punctis semicircumferentiae BLD motu illo semper aequabili, et continuo; cui nempe descriptioni correspondet descriptio illius lineae DKC a punctis diametraliter correlativis alterius semicircumferentiae BHD: obvium est intelligere, quod ipsa recta BA motu illo semper aequabili, et continuo describatur ab eo unico puncto B, quod nempe (veluti replicatum) intelligatur cum ipsa tali semicircumferentia semper excurrere super ea BA; dum interim eodem ipso tempore, motu eodem semper aequabili, et continuo, describitur illa altera DKC ab altero diametraliter correlativo unico puncto D, quod ipsum rursum (veluti replicatum) intelligatur cum sua altera semicircumferentia BHD semper excurrere super praedicta DKC. Tunc autem facilius intelligitur intenta aequalitas inter eam DKC, et eidem contrapositam rectam BA; quippe quae duae aequali ipso tempore, et aequali motu intelliguntur descriptae a duobus exactissime inter se aequalibus punctis, seu mavis infinitesimis. Ubi constat hanc ipsam exactissimam praedictorum punctorum aequalitatem non esse mihi in ista nova contemplatione necessariam. [95]

PROPOSITIO XXXVIII.

Hypothesis anguli acuti est absolute falsa, quia se ipsam destruit.

Demonstratur. Nam supra ex ipsa hypothesi anguli acuti evidenter elicuimus, curvam CKD (fig. 46.) ex ea

hypothesis of acute angle, is to be thought equal to the opposite base AB.

And that is what I had undertaken again to demonstrate by this new method. ✲

SCHOLION II. ✲

But again: since the straight BA is discerned as successively described by the points of the semicircumference BLD by that motion always uniform and continuous; to which description corresponds the description of that line DKC by the diametrically correlative points of the other semicircumference BHD: it is easy to understand, that this straight BA by that motion always uniform, and continuous is described by the one point B, which of course (as if unrolled) is thought always to run out with that semicircumference upon BA; whilst meanwhile in exactly the same time, by the same motion always uniform, and continuous, is described that other DKC by the other one diametrically correlative point D, which again itself (as if unrolled) is thought with its other semicircumference BHD always to run out upon the aforesaid DKC.

But then is more easily understood the asserted equality between DKC, and the straight BA opposite it; since the two are imagined to be described in equal time, and equal motion by two exactly equal points, or, if you prefer, infinitesimals. ✲

Where it holds that this exact equality of the aforesaid points is not necessary for me in that new consideration. [95]

PROPOSITION XXXVIII.

The hypothesis of acute angle is absolutely false, because it destroys itself.

PROOF. Assuredly we have above clearly deduced from the hypothesis of acute angle, that the curve CKD

prognatam aequalem esse debere contrapositae basi AB. Nunc autem contradictorium ex eadem hypothesi elicimus, quod curva CKD nequeat esse aequalis illi basi, cum certe sit eadem major. Quod enim curva CKD major sit recta CD ejus extremitates jungente, notio est omnibus communis, quam etiam demonstrare possumus ex vigesima primi, quod duo trianguli latera reliquo semper sunt majora; junctis nimirum CK, et KD; ac rursum junctis similiter apicibus, primo quidem duorum, tum quatuor, et sic in infinitum, duplicato numero enascentium segmentorum, quousque intelligatur hoc pacto absumi, seu desinere in ipsas infinite parvas seu chordas, seu tangentes, tota curva CKD. Sed hic procedere possumus ex sola communi notione. Quod autem juncta CD major sit basi AB, demonstratum a nobis est in 3. hujus ex ipsis visceribus hypothesis anguli acuti. Igitur curva CKD, ex hypothesi anguli acuti enascens, est certe major basi AB, quia est major, saltem ex communi notione, recta CD, quae ex hac ipsa hypothesi anguli acuti demonstratur major basi AB. Non igitur potest simul consistere, quod curva ista CKD aequalis sit basi AB. Itaque constat hypothesim anguli acuti esse absolute falsam, quia se ipsam destruit.

(fig. 46) born of it must be equal to the opposite base AB. But now we deduce the contradictory from the same hypothesis, that the curve CKD cannot be equal to that base, since surely it is greater than it.

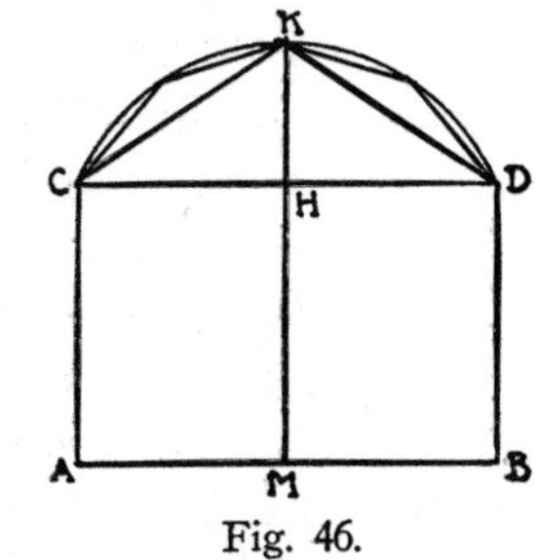

Fig. 46.

For that the curve CKD is greater than the straight CD joining its extremities, the notion is common to all, which also we may demonstrate from Eu. I. 20, that two sides of a triangle are always greater than the third; join CK, and KD; and again join likewise the apices, first of two, then of four, and so on *in infinitum,* the number of the produced segments doubling, until the whole curve CKD is understood in this way to be exhausted, or to end in those infinitely small chords, or tangents.

However here we may proceed from the common notion alone.

But that the join CD is greater than the base AB, has been demonstrated by us in P. III. from the very viscera of the hypothesis of acute angle.

Therefore the curve CKD, born of the hypothesis of acute angle, is certainly greater than the base AB, because it is greater, anyhow from the common notion, than the straight CD, which from the hypothesis of acute angle is demonstrated greater than the base AB. Therefore cannot at the same time stand, that the curve CKD is equal to the base AB.

Consequently is established that the hypothesis of acute angle is absolutely false, because it destroys itself.

SCHOLION.

Observare tamen debeo, quod etiam ex hypothesi anguli obtusi enascitur curva quaedam CKD, sed con-[96] vexa versus partes basis AB. Nam MH (fig. 47.) bifariam dividens ipsas AB, CD erit (ex 2. hujus) eisdem perpendicularis; et major (ex Cor. I. post 3. hujus) ipsis AC, BD, in hypothesi anguli obtusi. Quare ipsius MH portio quaedam MK aequalis erit ipsi AC, aut BD. Tum junctis CK, et KD, divisisque bifariam in punctis X, P, Q, N rectis CK, AM, MB, KD, constat (ex eadem 2. hujus) junctas PX, QN, perpendiculares fore ipsis rectis divisis. At rursum erunt illae (ex eodem Cor. I. post 3. hujus) majores ipsis AC, MK, BD. Hinc; assumptis earundem portionibus PL, QS, quae praedictis aequales sint; habebimus curvam, ex hypothesi anguli obtusi enascentem, quae transibit per puncta C, L, K, S, D. Atque ita semper, si decernere velimus reliqua puncta ejusdem curvae. Inde autem constat eam fore convexam versus partes basis AB. Jam fateor demonstrari uniformi plane methodo potuisse aequalitatem hujus curvae cum ipsa basi AB. At quis fructus? Nullus sane. Quemadmodum enim curva ista CKD censeri debet, ex communi saltem notione, major recta CD; ita etiam (in 3. hujus) basis AB demonstratur major eadem CD, dum stet hypothesis anguli obtusi. Nullum ergo ex hac parte absurdum, si basis AB aequalis sit praedictae curvae. Aliter vero rem

SCHOLION.

I should still observe, that also from the hypothesis of obtuse angle is born a certain curve CKD, but convex [96] toward the side of the base AB.

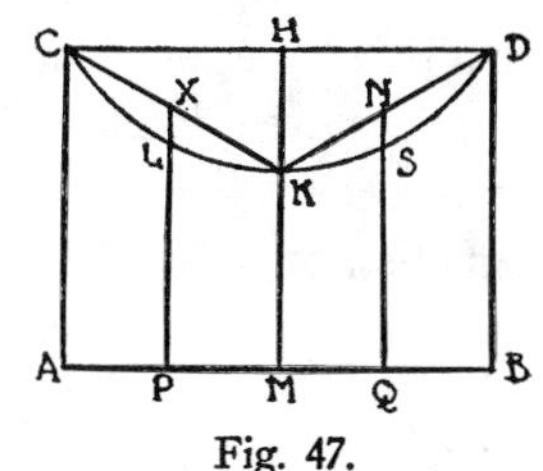

Fig. 47.

For MH (fig. 47) bisecting AB, CD will be (from P. II.) perpendicular to them; and greater (from Cor. I. to P. III.) than AC, BD, in the hypothesis of obtuse angle.

Wherefore a certain portion MK of MH will be equal to AC, or BD.

Then CK and KD being joined, and the straights CK, AM, MB, KD bisected in the points X, P, Q, N, it follows (again from P. II.) that the joins PX, QN will be perpendicular to the divided straights.

But again they will be (from the same Cor. I. to P. III.) greater than AC, MK, BD.

Hence; taking of them the portions PL, QS, which are equal to the aforesaid; we shall have a curve, born of the hypothesis of obtuse angle, which will go through the points C, L, K, S, D. And so on always, if we wish to determine remaining points of the same curve.

But thence follows it will be convex toward the side of the base AB. Now I grant in just the same way could have been demonstrated the equality of this curve with its base AB. But what good? None at all.

For just as the curve CKD must be thought, anyhow from the common notion, greater than the straight CD; so also (in P. III.) the base AB is proved greater than CD, when the hypothesis of obtuse angle holds. Therefore from this side is nothing absurd, if the base AB be equal to the aforesaid curve.

procedere in hypothesi anguli acuti, constat ex dictis supra.

Ex hoc igitur Scholio, et ex altero post 13. hujus intelligi potest, diversam plane viam iniri debuisse ad refellendam utranque falsam hypothesim, unam anguli obtusi, et alteram anguli acuti.

Praeterea facile itidem est ex istis dignoscere, non nisi rectam lineam CD esse posse, quae omnibus suis punctis aequidistet ab ea supposita recta linea AB. [97]

PROPOSITIO XXXIX.

Si in duas rectas lineas altera recta incidens, internos ad easdemque partes angulos duobus rectis minores faciat, duae illae rectae lineae in infinitum productae sibi mutuo incident ad eas partes, ubi sunt anguli duobus rectis minores.

Et hoc est notum illud Axioma Euclidaeum, quod tandem absolute demonstrandum suscipio. Ad hunc autem finem satis erit recolere nonnullas praecedentium Demonstrationum. Itaque in meis Propositionibus, usque ad VII. hujus inclusive, tres secrevi hypotheses circa rectam jungentem extrema puncta duorum aequalium perpendiculorum, quae uni cuidam rectae, quam basim appello, in eodem plano insistant. Porro circa has hypotheses (quas invicem secerno ex specie angulorum, qui ad eam jungentem fieri censeantur) demonstro unam quamlibet earum, nimirum aut anguli recti, aut anguli obtusi, aut anguli acuti, si vel in uno casu sit vera, semper et in omni casu illam solam esse veram. Tum in XIII. ostendo universalem veritatem Axiomatis controversi, dum consistat alterutra hypothesis aut anguli recti, aut anguli ob-

But that the thing goes otherwise in the hypothesis of acute angle, follows from what is said above.

From this scholion therefore and from the other after P. XIII. may be realized, that a wholly different way was to be taken in refuting each false hypothesis, one of obtuse angle, and the other of acute angle.

Moreover it is easy in like manner to recognize from these, that it can only be a straight line CD, which in all its points is equidistant from the assumed straight line AB. [97]

PROPOSITION XXXIX.

If upon two straight lines another straight striking makes toward the same parts angles less than two right angles, those two straight lines produced in infinitum meet each other toward those parts where are the angles less than two right angles.

This is the famous Euclidean axiom, which at length I undertake absolutely to demonstrate.

For this end however it will be sufficient to recall some of the preceding demonstrations. Therefore in my propositions, up to P. VII. inclusive, I have distinguished three hypotheses about the straight joining the extreme points of two equal perpendiculars, which stand upon a certain straight, that I call base, in the same plane.

Furthermore in regard to these hypotheses (which in turn I distinguish from the species of the angles, which are supposed to be made at the join) I demonstrate that any one of them, forsooth either of right angle, or obtuse angle, or acute angle, alone is true always and in every case, if even in one case it be true.

Then in P. XIII. I show the universal truth of the controverted axiom, when occurs either the hypothesis of right angle, or of obtuse angle.

tusi. Hinc in XIV. declaro absolutam falsitatem hypothesis anguli obtusi, quia se ipsam destruentis, utpote quae praedicti Axiomatis veritatem infert, ex quo contra reliquas duas hypotheses soli hypothesi anguli recti locus relinquitur. Igitur sola restat hypothesis anguli acuti, contra quam diutius pugnandum fuit.

Et hujus quidem (post multa, ne dicam omnia, conditionate expensa) absolutam falsitatem in XXXIII. tandem ostendo, quia repugnantis naturae lineae rectae, circa quam multa ibi intersero necessaria Lemmata. Tandem vero in praecedente Propositione absolute demonstro sibi ipsi repugnantem hypothesin anguli acuti. Quoniam igi-[98]tur unica restat hypothesis anguli recti, consequens plane est, ut ex praedicta XIII. hujus stabilitum absolute maneat praenunciatum Euclidaeum Axioma. Quod erat propositum.

SCHOLION.

Sed juvat expendere hoc loco notabile discrimen inter praemissas duarum hypothesium redargutiones. Nam circa hypothesin anguli obtusi res est meridiana luce clarior; quandoquidem ex ea assumpta ut vera demonstratur absoluta universalis veritas controversi Pronunciati Euclidaei, ex quo postea demonstratur absoluta falsitas ipsius talis hypothesis; prout constat ex XIII. et XIV. hujus. Contra vero non devenio ad probandam falsitatem alterius hypothesis, quae est anguli acuti, nisi prius demonstrando; quod linea, cujus omnia puncta aequidistent a quadam supposita recta linea in eodem cum ipsa plano existente, aequalis sit ipsi tali rectae; quod ipsum tamen non videor

Hence in P. XIV. I declare the absolute falsity of the hypothesis of obtuse angle, because it destroys itself, inasmuch as it occasions the truth of the aforesaid axiom, from which against the remaining two hypotheses place is left for the hypothesis of right angle alone. Therefore remains only the hypothesis of acute angle, against which was longer to be fought.

And of this indeed (after many things, I do not say all, circumstantially considered) at length in P. XXXIII. I show the absolute falsity, because repugnant to the nature of the straight line, about which I there introduce many necessary lemmata.

Finally in the preceding proposition I absolutely prove the hypothesis of acute angle contradictory to itself.

Since therefore [98] the hypothesis of right angle alone remains, the consequence plainly is, that from the aforesaid P. XIII. remains absolutely established the enunciated Euclidean axiom.

Quod erat propositum.

SCHOLION.

It is well to consider here a notable difference between the foregoing refutations of the two hypotheses. For in regard to the hypothesis of obtuse angle the thing is clearer than midday light; since from it assumed as true is demonstrated the absolute universal truth of the controverted Euclidean postulate, from which afterward is demonstrated the absolute falsity of this hypothesis; as is established from P. XIII. and P. XIV.

But on the contrary I do not attain to proving the falsity of the other hypothesis, that of acute angle, without previously proving; that the line, all of whose points are equidistant from an assumed straight line lying in the same plane with it, is equal to this straight, which itself finally I do not appear to demonstrate from the

demonstrare ex visceribus ipsiusmet hypothesis, prout opus foret ad perfectam redargutionem.

Respondeo autem triplici medio usum me fuisse in XXXVII. hujus ad demonstrandam praedictam aequalitatem. Et primo quidem, in corpore illius Propositionis, demonstro eam curvam CKD, prout enascentem ex hypothesi anguli acuti (ac propterea semper cavam versus partes illius rectae AB) aequalem eidem esse debere, et quidem argumentum ducendo ex ipsis ejusdem curvae tangentibus. Deinde in duobus ejusdem Propositionis subsequentibus Scholiis, praecisive a qualibet speciali hypothesi, bis rursum demonstro aequalitatem illius genitae lineae CD cum subjecta recta linea AB, qualiscunque tandem censeatur esse ipsa linea CD eo modo genita.

Jam vero; quatenus illa curva CKD, prout enascens [99] ex hypothesi anguli acuti, censeatur primo illo modo demonstrata aequalis subjectae rectae lineae AB; manifesta evadit redargutio, cum ex eadem hypothesi evidenter demonstretur major. Sin autem alterutro ex duobus aliis modis ostensa censeatur aequalitas praedicta; neque tunc cessat redargutio contra hypothesin anguli acuti. Ratio est; quia nihil vetat, quin illa CD sit curva, et nihilominus aequalis sit illi rectae AB, dum tamen sit semper versus eas partes convexa, ac propterea recta jungens illa eadem puncta C, et D minor sit contraposita basi AB, prout in hypothesi anguli obtusi: At omnino repugnat, si versus easdem partes sit semper cava, ac propterea recta jungens praedicta illa puncta C, et D major sit eadem contraposita basi AB, prout in hypothesi anguli acuti. Atque ita declaratum jam est in Scholio praecedentis Propositionis. Scilicet contra hypothesin anguli obtusi

viscera of the very hypothesis, as must be done for a perfect refutation.

But I reply I used a triple means in P. XXXVII. for demonstrating the mentioned equality.

And first, in the body of the proposition, I prove the curve CKD, as born from the hypothesis of acute angle (and therefore always concave toward the side of the straight AB) must be equal to it, and indeed by drawing the argument from the tangents of the curve.

Then in two subsequent scholia of the proposition, apart from any special hypothesis, twice again I demonstrate the equality of the generated line CD with the underlying straight line AB, of whatever kind the line CD so generated is supposed to be.

But now; in so far as the curve CKD, as born [99] from the hypothesis of acute angle, is judged to be proved by the first method equal to the underlying straight line AB, a manifest refutation arises, since from the same hypothesis it is evidently proved greater. But if the aforesaid equality is supposed shown in either of the two other modes; not even then does the refutation cease against the hypothesis of acute angle. The reason is; because nothing forbids, that CD may be curved, and nevertheless may be equal to the straight AB, while yet it may be always convex toward that side, and therefore the straight joining the points C, and D may be less than the opposite base AB, as in the hypothesis of obtuse angle. But, it is wholly contradictory, if toward that side it be always concave, and therefore the straight joining the points C, and D be greater than the opposite base AB, as in the hypothesis of acute angle.

And so has just now been stated in the scholion of the preceding proposition.

Of course against the hypothesis of obtuse angle it

manifestum est nullam hinc sequi redargutionem, quae propterea unice impetit hypothesin anguli acuti.

Hoc tamen loco aliquis fortasse inquiret, cur adeo sollicitus sim in demonstranda utriusque falsae hypothesis exacta redargutione. Ad eum, inquam, finem, ut inde magis constet non sine causa assumptum fuisse ab Euclide tanquam per se notum celebre illud Axioma. Nam hic maxime videtur esse cujusque primae veritatis veluti character, ut non nisi exquisita aliqua redargutione, ex suo ipsius contradictorio, assumpto ut vero, illa ipsa sibi tandem restitui possit. Atque ita a prima usque aetate mihi feliciter contigisse circa examen primarum quarundam veritatum profiteri possum, prout constat ex mea Logica demonstrativa.

Inde autem transire possum ad explicandum, cur in Proemio ad Lectorem dixerim: *non sine magno in rigidam Logicam peccato assumptas a quibusdam fuisse tanquam datas* [100] *duas rectas lineas aequidistantes.* Ubi monere debeo nullum eorum a me hic carpi, quos in hoc meo Libro vel indirecte nominavi, quia vere magnos Geometras, hujusque peccati verissime immunes. Dico autem: *magnum in rigidam Logicam peccatum*: quid enim aliud est assumere tanquam datas *duas rectas lineas aequidistantes*: nisi aut velle; quod omnis linea in eodem plano aequidistans a quadam supposita linea recta sit ipsa etiam linea recta; aut saltem supponere, quod una aliqua sic aequidistans possit esse linea recta, quam idcirco seu per hypothesin, seu per postulatum praesumere liceat in tanta aliqua unius ab altera distantia? At constat neutrum horum venditari posse tanquam per se notum. Scilicet ratio objectiva lineae, quae omnibus suis punctis aequidistet a quadam supposita linea recta, non ita clare per

is manifest no refutation follows hence, which therefore only demolishes the hypothesis of acute angle.

In this place however some one perchance may inquire, why I am so solicitous about proving exact the refutation of each false hypothesis. To the end, say I, that thence may be more completely established that not without cause was that famous axiom assumed by Euclid as known *per se.* For chiefly this seems to be as it were the character of every primal verity, that precisely by a certain recondite argumentation based upon its very contradictory, assumed as true, it can be at length brought back to its own self. And I can avow that thus it has turned out happily for me right on from early youth in reference to the consideration of certain primal verities, as is known from my *Logica demonstrativa.*

Thence now I may proceed to explain, why in the Preface to the Reader I have said: *not without a great sin against rigid logic two equidistant straight lines have been assumed by some as given.* [100]

Where I should point out that none of those is carped at, whom I have mentioned even indirectly in this book of mine, because they are truly great geometers, and verily free from this sin.

But I say: *great sin against rigid logic*: for what else is it to assume as given *two equidistant straight lines*: unless either to assume; that every line equidistant in the same plane from a certain supposed straight line is itself also a straight line; or at least to suppose, that some one thus equidistant may be a straight line, as if therefore it were allowable to make assumption, whether by hypothesis, or by postulate, of any such distance of one from another? But it is certain neither of these can be made traffic of as if *per se* known.

Forsooth the objective concept of a line, which in all its points is equidistant from a certain supposed straight

se ipsam congruit cum definitione propria ipsius lineae rectae. Quare assumere duas rectas lineas sub ista *aequidistantiae* ratione inter se *parallelas* fallacia est, quam in praedicta mea Logica appello *Definitionis complexae,* juxta quam irritus est omnis progressus ad assequendam veritatem absolute talem.

Unam tamen superesse adhuc video necessariam observationem. Nam lineam jungentem extrema puncta omnium aequalium perpendiculorum, quae in eodem plano versus easdem partes erigantur a singulis punctis subjectae rectae lineae AB, debere esse et aequalem praedictae AB, et rursum in seipsa rectam, fateri omnes volumus. Sed dico prius esse apud nos, quod aequalis sit; deinde autem, quod recta. Cum enim singula puncta illius rectae AB intelligi possint semper aequabiliter procedere per sua illa perpendicula ad formandam tandem illam qualemcunque CD; manifestum videri debet, quod illa qualiscunque genita CD aequalis sit eidem AB; praesertim vero, si respiciamus explicationem contentam in Scholio II. post [101] XXXVII. hujus, ubi hoc punctum clarissime demonstratum est.

Sed postea magna adhuc restat difficultas in demonstrando, quod illa eadem sic genita CD non nisi recta linea sit. Atque hinc factum esse puto, ut ex communi quadam persuasione rectam lineam, pro faciliore progressu, maluerint praesumere, ut inde aequalem ostenderent illi basi AB, ac postea inferrent rectos angulos ad ipsam talem jungentem CD. Dico autem *magnam difficultatem*: Nam prius expendere oportebat tres hypotheses circa angulos ad illam junctam *rectam* CD, nimirum aut rectos, si ipsa aequalis sit basi AB; aut obtusos, si minor;

line, clearly is not thus *per se* congruent with the proper definition of the straight line.

Wherefore to define two *parallel* straight lines under this relation of mutual *equidistance* is the fallacy, which in my aforesaid *Logica* I call *definitionis complexae,* in connection with which every advance toward attaining truth absolutely such is ineffectual.

I see in addition there still remains one necessary observation.

For we all are willing to grant the line joining the extreme points of all equal perpendiculars, which in the same plane are erected toward the same parts from the separate points of an underlying straight line AB, must be both equal to the aforesaid AB, and moreover in itself straight.

But I say with us is first, that it is equal; then however, that it is straight.

For since the single points of the straight AB may be thought always to proceed uniformly upon those perpendiculars of theirs to forming at length that certain CD; it should seem manifest, that the generated CD, of whatsoever kind, is equal to AB; but especially, if we consider the explication contained in Scholion II. after [101] P. XXXVII., where this point is most clearly demonstrated.

But thereafter still remains a great difficulty in demonstrating, that this same generated CD cannot be anything but a straight line. And hence comes it I think, that from a certain common conviction, for more facile progress, they have preferred to presume the line straight, that thence they might show it equal to the base AB, and afterward infer right angles at the join CD.

But I say *great difficulty*: For first it was necessary to consider three hypotheses about the angles at the *straight* join CD, forsooth either right, if it be equal to the base AB; or obtuse, if less; or acute, if greater. But

aut acutos, si major. Tum vero ostendi debebat non nisi cavam esse posse versus basim AB lineam curvam, quae (in hypothesi anguli acuti) jungat extremitates illorum aequalium perpendiculorum, ac rursum non nisi convexam versus eandem basim aliam curvam, quae (in hypothesi anguli obtusi) jungat extremitates eorundem perpendiculorum. Deinde autem hypothesis quidem anguli acuti ex eo demonstranda erat falsa; quia linea jungens praedictorum perpendiculorum extremitates adeo non erit aequalis basi AB, ut immo (ex communi saltem notione) major sit illa juncta recta CD, quae ex natura ipsiusmet hypothesis major est praedicta basi AB. At hypothesis anguli obtusi aliunde ostendenda erat sibi ipsi repugnans, prout in XIV. hujus. Sed haec jam satis.

Finis Libri primi.

then it had to be shown that the curved line, which (in the hypothesis of acute angle) joins the extremities of those equal perpendiculars, could only be concave toward the base AB, and again the other curve, which (in the hypothesis of obtuse angle) joins the extremities of the same perpendiculars, only convex toward the same base. But then the hypothesis indeed of acute angle from this was demonstrated false; because the line joining the extremities of the aforesaid perpendiculars was so far not equal to the base AB, as on the contrary (anyhow from the common notion) it is greater than the straight join CD, which from the nature of this hypothesis itself is greater than the aforesaid base AB.

But the hypothesis of obtuse angle had to be shown from another source contradictory to itself, as in P. XIV.

But this now is enough.

End of Book I.

NOTES.

Page 21. Prop. III: Euclid's first two postulates are: Let it be granted,

1. that one and only one sect can be drawn from any point to any other;

2. and that this sect may be produced continually on its straight.

In I. 16 he assumes that the straight divides the plane into two separate regions, and also the Archimedes assumption that the straight is infinite and open. This block of assumptions is incompatible with the hypothesis of obtuse angle as Saccheri later shows. If it were also incompatible with the hypothesis of acute angle, we should have a perfect case of Saccheri's favorite method. The proofs would be fairy proofs leading to a direct demonstration of their contradictory opposite; and none of them could make part of a modern treatise on non-Euclidean geometry.

But since Euclid's assumptions, barring the Parallel Postulate, are perfectly compatible with the hypothesis of acute angle, many of Saccheri's proofs remain the most elegant and cogent the world possesses in the domain of non-Euclidean geometry.

Page 27. Prop. III, Cor. II: Saccheri simply cites this corollary when, as often, he wishes the proposition: In *any* birectangular quadrilateral HMPC with angle P obtuse and angle C acute, side PM $<$ CH.

To this the proof of Prop. III, Part 3, applies.

PAGE 33. Prop. VI: Here is assumed the *principle of continuity.* An elementary proof without this and without the Archimedes assumption is given by Bonola.

Without these, and with only a sect-carrier replacing the circle in constructions, Euclid's geometry and a geometry fulfilling the hypothesis of obtuse angle are given in Halsted, *Géométrie rationnelle,* Paris, Gauthier-Villars. Compare, for the hypothesis of acute angle: John Bolyai, *The Science Absolute of Space,* translated from the Latin by Dr. George Bruce Halsted; and Nicholas Lobachevski, *Geometrical Researches on the Theory of Parallels,* translated from the original by George Bruce Halsted (The Open Court Publishing Company, 1914).

PAGE 95f. Demonstrations physico-geometric. If in a single case the angle inscribed in a semicircle be ascertained to be right, Euclidean geometry is established. But measurement being imperfect, this is hopeless. What if such angle were found other than right by a difference greater than the limits of experimental error?

Consult: George Bruce Halsted, *The Foundations of Science,* New York, The Science Press, 1913.

PAGE 109. Prop. XXI. Scholion IV: Saccheri misses the possibility that the intersection point P of APY and XPY may go to infinity while AX remains finite.

PAGE 192. *Ly* is a term of the grammarians and rhetoricians, by which is denoted the treatment of a word as itself a thing. The Greek article τὸ was thus used.

PAGE 221. Prop. XXXVII. Scholion I: Is it possible Saccheri did not perceive that his reasoning applies just as well to proving two concentric circles equal?

PAGE 224. Prop. XXXVII. Scholion II: "aequali tempore," yes; but "aequali motu" is unproven.

EDITOR'S NOTES

Page 5, line 1. Pages iii-iv of the original Latin text contain the Dedication; page vii, the permission to print (*imprimatur*) of the provincial superior in Milan, of August 16, 1733; page viii, the *imprimatur* of the Inquisition, of July 13, 1733.

Page 5, line 2. *Alternatively:* No one who has learned mathematics can fail to be aware of the extraordinary merit of Euclid's *Elements.* I call as expert witnesses Archimedes, Apollonius, Theodosius, and the almost innumerable other writers on mathematics up to the present who make use of Euclid's *Elements* as a long established and unshakeable foundation. But this great prestige of the *Elements* has not prevented many ancient as well as modern geometers, including many of the most distinguished, from claiming that they had found certain blemishes in this beautiful work, which cannot be too highly praised. Three such blemishes have been cited, which I now give.

The first has to do with the definition of parallels and, in connection with this, the axiom which in Clavius is the thirteenth of the First Book, where Euclid says:

If on two straight lines lying in the same plane there falls a straight line that makes the interior angles on the same side less than two right angles, then the two straight lines, if produced indefinitely, meet on that side on which the angles are less than two right angles.

No one doubts the truth of this proposition; rather, Euclid is blamed for calling it an axiom—as though a correct understanding of its wording made it self-evident. So

that not a few have tried to prove it — while retaining Euclid's definition of parallels — solely on the basis of those propositions of Euclid's First Book that precede the twenty-ninth, where the use of the controversial proposition begins.

Page 21, line 21. See Professor Halsted's Note, on page 242.

Page 23, line 20. It would be better to put it this way: *for the angles at* A *and at* C *are not equal.*

Page 25, line 4. The Proposition on Exterior Angles (Eu. I. 16) that is being used here presupposes that the straight line is of *infinite length*. An alternative assumption, suggested by Riemann in his memoir *Ueber die Hypothesen welche der Geometrie zugrunde liegen* is that the straight line is *unbounded.* This property is compatible with the hypothesis that the straight line is infinite (open) as well as with the hypothesis that it is finite (closed). If the first of these two hypotheses is taken, then, as Saccheri shows, the hypothesis of the obtuse angle leads to a contradiction. If the straight line is, on the other hand, taken to be (merely) *unbounded*, which allows the second possibility — that the straight line is finite (closed) — then the Proposition on Exterior Angles (Eu. I. 16) cannot be invoked, and Saccheri's proof becomes insufficient.

Indeed, if this second possibility is taken, if the straight line is assumed to be a closed line of finite length, then a contradiction can be shown *not* to occur. (See Bonola, *Non-Euclidean Geometry* [2], pages 141 and 142.)

Page 27, line 16. Corollary II of Proposition III will later be used a number of times in the following form: For *every* quadrilateral HCPM having right angles at H and at M and having an acute angle at C and an obtuse angle at P, side PM must be smaller than side CH. This is not immediately clear from the wording of the corollary; but the above Proof of the Third Part may indeed be applied to *any and every* such quadrilateral HCPM. See also Profes-

sor Halstead's Note (to the same effect) on page 242.

Page 31, line 11. Saccheri calls the second proof more elegant because it is strictly Euclidean. But in truth it likewise requires a reflection in the base CD, since ACD must be applied upon RCD, and thus the intrinsic difficulty remains.

Page 31, line 13. Note that it is presented here as self-evident that the straight line is of infinite length. See also the Editor's Note to page 25, line 4.

Page 35, line 1. Here it is tacitly assumed that the length of the line changes *continuously* in the transition from CD to RX. But, as was shown by Lambert, the result is independent of that assumption. See also Professor Halsted's Note on page 243.

Page 41, line 12. Saccheri, in saying *the two acute angles remaining,* is using Eu. I. 17, to the effect that the sum of any two angles of a triangle is less than two right angles. However, under the hypothesis of the obtuse angle, Eu. I. 17 is not valid, since it is obtained as an immediate consequence of Eu. I. 16, the Proposition on the Exterior Angle. And indeed, Saccheri later proves, in Proposition XIV, that *the hypothesis of the obtuse angle is self-cancelling,* since it leads to a contradiction against Eu. I. 17. See also the Editor's Note to page 25, line 4.

Page 49, line 16. Proposition XII is correct, but the proof that follows lacks cogency, since it makes use of Eu. I. 16, the Proposition on the Exterior Angle, which is not valid under the hypothesis of the obtuse angle.

Page 53, line 1. The comment made just above concerning Proposition XII applies equally to Proposition XIII.

Page 55, line 17. Thus we would have a case where the straights AD and XL do not meet even though the sum of the interior angles LXA and XAD is smaller than two right angles.

Page 55, line 10 from below. The sense of the following remarks is to this effect: Given two angles that add up to less than two right angles, we can always construct triangles in which these angles occur. Therefore if we choose the side adjacent to the two angles of such a triangle as our base AX, then we have obtained the desired intersection property for the two angles. What is left undecided, however, is the question of whether *any given* base AX can be obtained in this manner, which after all would be necessary for a complete proof of the Euclidean axiom.

Page 57, line 14. In many diagrams, Saccheri uses the *same* letter—in this case, X—to designate *different* points that in certain respects, however, are on an equal footing. This usage is not just peculiar to Saccheri but is often found in the older mathematical literature—for instance, in the geometric researches of Blaise Pascal (1623–1660) [*Oeuvres complètes,* t. III, Paris, 1882, pp. 370–446], or in the *Lectiones Geometricae,* London, 1670 (2nd ed., London, 1674) of Isaac Barrow (1630–1677) [*The Geometrical Lectures of Isaac Barrow,* Chicago and London, 1916], or in John Bernoulli's (1667–1748) paper on the brachistochrone (*Acta Eruditorum,* May, 1697). The number of such examples could be multiplied at will.

The use of subscripts—an eminently convenient device, dating back to suggestions made by Leibniz toward the end of the seventeenth century—did not become common practice among mathematicians before the nineteenth century.

Page 57, line 16. Saccheri is not bothered by the fact that angle K in his Fig. 12 is about 30° rather than 84°, or that angle R is about 120° rather than 91°, since clearly he only took a *schematic* view of his figures, as will be even more obvious later on; note, for instance, the *right* angles in Fig. 19 below (page 79). In this translation, the diagrams are reproduced in their original form, without any

attempt to reconcile diagrams and text.

Page 59, line 9. Saccheri here assumes that angle BXA varies *continuously* as point X travels from A to P. See also the Editor's Note to page 35, line 1.

Page 59, line 20. To wit, below in Proposition XVII and its first scholion.

Page 61, line 20. What has been said above regarding Propositions IX and XII (Editor's Notes to page 41, line 12 and page 49, line 16) applies also to the proof of Proposition XV given here.

Page 81, line 6. Saccheri deliberately says *not greater,* since Proposition XX, as given, is also valid under the hypothesis of the right angle.

Page 81, line 12. *The same remaining.* I.e., under the same hypotheses as before.

Page 87, line 17. Euclid, *Elements,* Book I, Definition 23.

Page 87, line 8 from below. Euclid, *Elements,* Book I, Definition 22.

Page 89, line 10 from below. This diagram already occurs in Clavius, 1574 and in Giordano da Bitonto, 1680.

Page 93, line 5. What is meant by *superposition* (*superpositio* in the Latin original) is a reflection of the figure in the straight FG. See also the Editor's Note to page 31, line 11.

Page 93, line 3 from below. *Alternatively:* Therefore, in a matter such as this, one cannot be blamed for making a somewhat greater effort to get at the exact truth.

Page 95, line 6. See Professor Halsted's note, on page 243.

Page 99, line 2 from below. Saccheri interprets Euclid's definition of a straight line in a way foreign to Euclid's intent, since Euclid carefully avoids the concept of motion.

Page 109, line 2. Saccheri's use of Eu. I. 16, and thus his exclusion of the hypothesis of the obtuse angle, is an essential *defect* in his proof. For, the assumption that there

exist two similar triangles is already sufficient to invalidate *both* hypotheses, that of the obtuse angle as well as that of the acute angle. Consult, for example, Lambert's *Theorie der Parallellinien,* §§79 and 80, or Bonola, *Non-Euclidean Geometry,* [2], page 29, footnote.

Page 109, line 5 from below. Viz, in scholion II.

Page 109, line 2 from below. APY being regarded, as the sequel shows, as the extension of the straight line AB.

Page 113, line 2. The discussion here has overlooked the possibility of point P being at infinity, in which case no contradiction would result.

Page 115, line 3 from below. Here, too, Saccheri makes use of the axiom of continuous variation; see also the Editor's Note to page 35, line 1, concerning the proof of Proposition VI.

Page 117, line 10. Saccheri here uses one and the same letter X twice—probably because, in the case where the straights AD and BK meet, he thinks of X as their point of intersection.

Page 125, line 2. *Alternatively:* Under the same hypotheses. That is, assuming, as in the proof of Proposition XXIII above, the case in which all the angles ADK, AHK, etc. are obtuse.

Page 129, line 12. Saccheri means to say that the straights AX and BX are assumed only to meet at infinity. Without this assumption one could not claim, in the proof that follows, that MN must be less than AB, for M could then lie beyond the point of intersection of the two lines.

Page 139, line 12. To wit, in Corollary I to Proposition XXVII.

Page 143, line 11. Saccheri here treats the intersection at infinity as if this were a finite point. His later proof of Euclid's Axiom (Proposition XXXIII below) is based on the same erroneous treatment.

Page 161, line 3. The meaning is *toward* AX *without any limitation, inside the angle* BAX.

Page 161, line 7 from below. For, Saccheri considers only the rays emanating from the point A. Then the *rays* that lie beyond the perpendicular AY, such as AZ, have no common perpendicular with BX.

Page 173, line 9 from below. See the Editor's Note to page 143, line 11.

Page 173, line 5 from below. *Alternatively:* I shall take the utmost care not to pass over any objection, however pedantic it might seem, since it appears to me that this is appropriate to a highly rigorous proof.

Page 177, line 10. Euclid, *Elements,* Book I, Postulate 1.

Page 183, line 5. Saccheri here envisions AX rigidly connected with XB, and this rigid system rotated about AX.

Page 183, line 18. Namely, in the new position AXC symmetric to AXB and generated by rotation about AX of the rigid system AXB.

Page 193, line 20. See Professor Halsted's Note to page 192, on page 243. See also Micraelius, J. *Lexicon Philosophicum,* Jena, 1653, page 608.

Page 201, line 1 from below. Euclid, *Elements,* Book I, Definition 10.

Page 203, line 12. Euclid expressly requires, in Postulate 4 of Book I, that all right angles be equal. His motivation in introducing this postulate was, presumably, to avoid the concept of motion.

Page 207, line 2. Cf. Proposition XXXIII above (page 173) and the Editor's Note to page 143, line 11.

Page 221, line 13. See Professor Halsted's Note, on page 243.

Page 221, line 16. Thus, Saccheri himself seems to have sensed that the proof just given is insufficient.

Page 223, line 13. If a straight line is tangent to a circle and if we draw a perpendicular to this tangent line through the point of contact, then the center of the circle lies on this perpendicular.

Page 229, line 4. This proof suffers from exactly the same insufficiency as the preceding proof.

Page 225, line 4. This proof suffers from exactly the same insufficiency as the preceding proof.

Page 225, line 5. See Professor Halsted's Note, on page 243.

Page 225, line 9 from below. This argument, like the preceding ones, are on shaky ground. True, the circle BHDL rolls along the staight line AB and thus unwinds from it; but it does not roll simultaneously on the curve DKC and therefore does not unwind from that curve.

INDEX

SUBJECT INDEX.

GEOMETRY

BAKER, H. F.: Introduction to Plane Geometry, with many Examples, 2nd. ed. 382 pp. 5⅜ x 8. ISBN -0247-7, ∞G $18.50

BIRKHOFF, George D. & BEATLEY R.: Basic Geometry, 3rd ed. 294 pp. 5⅜ x 8. ISBN -0120-9, ∞$12.00; Teacher's Manual, $2.50; Answer Book, $1.50

BLUMENTHAL, Leonard M.: Theory and Applications of Distance Geometry, 2nd ed. 359 pp. 5⅜ x 8. ISBN -0242-6, ∞G $16.95

BONNESEN, Tommy & FENCHEL, W.: Theorie der Konvexen Koerper. (Germ.) 171 pp. 5½ x 8½. ISBN -0054-7, ∞G $6.95

COOLIDGE, Julian L.: A Treatise on the Circle and the Sphere, 2nd (corr.) ed. 602 pp. 5⅜ x 8. ISBN -0236-1, ∞G $27.50

DARBOUX, Gaston: Theorie Generale des Surfaces, 3rd ed. 4 vols. (French) 2,283 pp. 5⅜ x 8. ISBN -0216-7, Four vol. set, ∞G $85.00

FORDER, Henry G.: The Calculus of Extension, xvi + 490 pp. 5⅜ x 8. ISBN -0135-7, ∞G $25.00

GOMES TEIXEIRA, Francisco: Traite des Courbes Speciales Planes et Gauches, 2nd ed. 3 vols. (French) 1,337 pp. 6½ x 10. ISBN -0255-8, ∞G $65.00 the set.

GRASSMANN, Hermann G.: Die Ausdehnungslehre von 1844/1878, 4th ed. (Germ.) xii + 435 pp. 6 x 9. ISBN -0222-1, ∞G $27.50

HILBERT, David: & COHN-VOSSEN, S.: Geometry and the Imagination. 358 pp. 6 x 9. ISBN -0087-3, ∞G $16.95

JESSOP, Charles H.: Treatise on the Line Complex. xv + 364 pp. 5⅜ x 8. ISBN -0223-X, ∞G $14.95

KLEIN, Felix: Hoehere Geometrie, 3rd ed. (Germ.) vi + 405 pp. 5⅜ x 8. ISBN -0065-2, G $9.95

———Nicht-Euklidische Geometrie. (Germ.) xii + 326 pp. 5⅜ x 8. ISBN -0129-2, ∞G $9.95

———Famous Problems of Elementary Geometry. See: KLEIN, et al.

———et al. Famous Problems and Other Monographs, 4 vols. in 1. *Contains*: KLEIN, Famous Problems of Elementary Geometry; SHEPPARD, From Determinant to Tensor; MACMAHON, Intro. to Combinatory Analysis, MORDELL, L. J., Three Lectures on Fermat's Last Theorem. 350 pp. 5⅜ x 8. ISBN -108-X, ∞ $11.95

LAGUERRE, Edmond: Oeuvres (Collected Mathematical Works), 2 vols. (French) 1,202 pp. 5⅜ x 8. ISBN -0263-9, ∞G $49.50 the set

LIE, Sophus: Geometrie der Beruehrungstransformationen. (Germ.) 705 pp. 6 x 9. CIP. ISBN -0291-4, ∞G $29.50

SACCHERI, Girolamo: Euclides Vindicatus, 2nd ed. (translated by G. B. Halsted and with added notes by P. Staeckel and F. Engels), xxx + 250 pp. 5⅜ x 8. ISBN -0289-2, ∞G

SALMON, George: Analytic Geometry of Three Dimensions. 7th ed. Vol. I. xxiv + 470 pp. 5⅜ x 8. ISBN -0122-5, O.P.; 5th ed. Vol. II. xvi + 334 pp. 5⅜ x 8. ISBN -0196-9, ∞G $12.95

———Conic Sections, 6th ed. 415 pp. 5⅜ x 8. ISBN -0099-7, ∞G $9.95, cl., ISBN -0098-9, $6.95, pa.

SCOTT, Charlotte A.: Projective Methods in Plane Analytic Geometry. 3rd ed. xii + 290 pp. 5⅜ x 8. ISBN -0146-2, ∞G $11.95

SNYDER, Virgil, et al: Selected Topics in Algebraic Geometry, 2 vols. in 1. 490 pp. 6 x 9. ISBN -0189-6, ∞G $15.95

SOMMERVILLE, Duncan M.Y.: Bibliography of Non-Euclidean Geometry, 2nd ed. xii + 410 pp. 5⅜ x 8. ISBN -0175-6, ∞G $22.50

STEINER, Jacob: Gesammelte Werke, 2 vols. (Germ.) 1,336 pp. 6 x 9. ISBN -0233-7, ∞G $59.50 the set

STERNBERG, Shlomo: Lectures on Differential Geometry. 2nd (corrected and supplemented) ed. xiii + 421 pp. 6 x 9. ISBN -0316-3, ∞G $25.00

YOUNG, Grace Chisholm and William Henry YOUNG: Beginner's Book of Geometry. 2nd ed. xvi + 222 pp. 4⅝ x 6⅜. ISBN -0231-0, ∞ $12.00

ANALYSIS

BANACH, Stefan: Theorie des Operations Lineaires. (French) xii + 250 pp. 5⅜ x 8. ISBN -0110-1, ∞G $10.95

BERBERIAN, Sterling K.: Introduction to Hilbert Space, 2nd ed. vi + 206 pp. 5⅜ x 8. CIP. ISBN -0287-6, ∞G $9.95

BOCHNER, Salomon: Fouriersche Integrale. (Germ.) vi + 229 pp. 5½ x 8½. ISBN -0042-3, G $10.50

BOHR, Harald: Almost Periodic Functions. 120 pp. 6 x 9. (lithotyped). ISBN -0027-X, ∞G $9.95

BOLZA, Oskar: Lectures on the Calculus of Variations, 3rd ed. 280 pp. 5⅜ x 8. ISBN -0145-4, ∞G $12.95

———Vorlesungen ueber Variationsrechnung. (Germ.) ix + 715 pp. 5⅜ x 8. ISBN -0160-8, ∞G $23.95

CARATHEODORY, Constantin: Calculus of Variations and Partial Differential Equations, 2nd (revised) ed. xvi + 401 pp. 6 x 9. ISBN -0318-X, ∞G $25.00

CORDUNEANU, Constantin: Principles of Differential and Integral Equations, 2nd ed., x + 205 pp. 6 x 9. ISBN -0295-7, T.O.P.

EDWARDS, Joseph: The Integral Calculus, 2 vols. 1,922 pp. 5 x 8. Vol. I, ISBN -0102-0, ∞G $45.00; Vol. II, ISBN -0105-5, ∞G $45.00

EVANS, Griffith C.: Logarithmic Potential and Other Monographs. Logarithmic Potential, 2nd ed.; & BLISS, G. A. Fundamental Existence Theorems, & KASNER, E. Differential-Geometric Aspects of Dynamics. 3 vols in 1. 399 pp. 5⅜ x 8. ISBN-0305-8, ∞G $19.50.

FORD, Lester R.: Automorphic Functions. 343 pp. 5⅜ x 8. ISBN -0085-7, ∞G $18.50

FORD, Walter B.: Studies in Divergent Series and Summability; Asymptotic Developments of Functions. 371 pp. 6 x 9. ISBN -0143-8, ∞G $19.50

GAMELIN, Theodore W.: Uniform Algebras, 2nd ed. xiii + 263 pp. 6 x 9. ISBN -0311-2, ∞G $17.95

GELFAND, Israel M., RAIKOV, D. A. & SHILOV, G. E.: Commutative Normed Rings. 306 pp. 6 x 9. ISBN -0170-5, ∞G $14.95

GOFFMAN, Caspar & PEDRICK, George: A First Course in Functional Analysis, 2nd. ed. xiv + 284 pp. 6 x 9. ISBN -0319-8, ∞G $15.95

GUNTHER, N.: Integrales de Stieltjes (French) iv + 494 pp. 5⅜ x 8. ISBN -0063-6, G $19.95

HALMOS, Paul R.: Introduction to Hilbert Space. 120 pp. 6 x 9. ISBN -0082-2, ∞G $9.95

———Lectures on Ergodic Theory, viii + 101 pp. 5⅜ x 8. ISBN -0142-X, ∞G $8.95

HOBSON, Ernest W.: Spherical and Ellipsoidal Harmonics. xi + 500 pp. 5⅜ x 8. ISBN -0104-7, ∞G $17.95

KRAZER, Adolf: Lehrbuch der Thetafunktionen. (Germ.) xxiv + 509 pp. 5⅜ x 8. ISBN -0244-2, ∞G $25.00

LANDAU, Edmund: Differential and Integral Calculus, 372 pp. 6 x 9. ISBN -0078-4, ∞G $16.95

———Foundations of Analysis, xiv + 136 pp. 6 x 9. ISBN -0079-2, ∞G $10.95

———Grundlagen der Analysis, with a complete German-English Vocabulary, 4th ed. 173 pp. 5½ x 8½. ISBN -0141-1, $4.95, pa.

LEBESGUE, Henri: Lecons sur L'Integration et la Recherche des Fonctions Primitives, 3rd ed. (French) xii + 340 pp. 5⅜ x 8. CIP. ISBN -0267-1 ∞G $13.95